Roger Haines on Report Writing

A Guide for Engineers

Roger Haines on Report Writing

A Guide for Engineers

Roger W. Haines

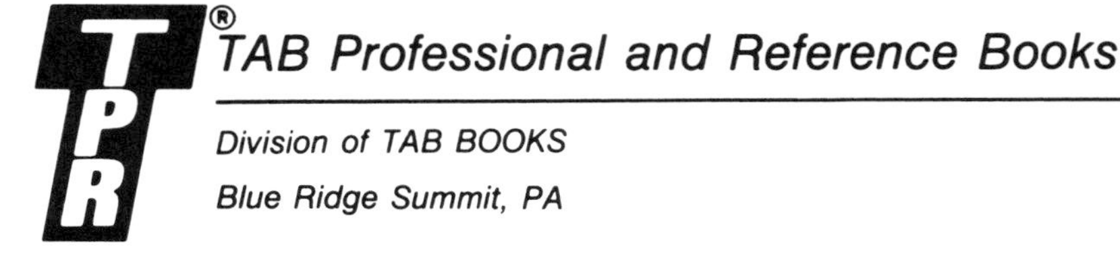

TPR books are published by TAB Professional and Reference Books, a division of TAB BOOKS. The TPR logo, consisting of the letters "TPR" within a large "T," is a registered trademark of TAB BOOKS.

Published by **TAB BOOKS**
FIRST EDITION/FIRST PRINTING

Library of Congress Cataloging-in-Publication Data

Haines, Roger W.
Roger Haines on report writing : a guide for engineers / by Roger W. Haines
p. cm.
ISBN 0-8306-3313-8 — ISBN 0-8306-4313-3 (pbk.)
1. Communication of technical information. I. Title.
T10.5.H35 1990 90-31300
620'.0014—dc20 CIP

TAB BOOKS offers software for sale. For information and a catalog, please contact TAB Software Department, Blue Ridge Summit, PA 17294-0850.

Questions regarding the content of this book should be addressed to:

Reader Inquiry Branch
TAB BOOKS
Blue Ridge Summit, PA 17294-0850

Vice President & Editorial Director: Larry Hager
Book Editor: Barbara B. Minich
Production: Katherine Brown
Book Design: Jaclyn J. Boone

Contents

Acknowledgments

This book could not have been written without the liberal education I received from Donald Bahnfleth. Don was the editor of *Heating/Piping/Air Conditioning* magazine when I first started writing for publication. He gave me much encouragement and many pointers. Later I worked for him for over four years. Using his prerogatives as the boss, he made me write and rewrite until he was satisfied, in the process honing my writing skills.

I would also like to recognize the basic training provided by Iowa State University, in the form of Freshman English classes taught by Dr. Leonard Feinberg, which consisted largely of writing (with sometimes severe criticism from classmates and instructor), and a senior course called "Writing Scientific Papers."

And, as always, my wife, Wilma, has been a constant source of help. In this book, I called on her expertise in grammar and syntax, where she made a number of contributions.

Introduction

Every engineer writes reports. Many are simple, informal letter reports. Some are elaborate, formal reports with masses of substantiating data. All reports, if they are to be credible, must follow certain forms and obey the rules of logic. The purpose of this book is to define, analyze, and explain the elements that must be present in any effective engineering report.

The axiom "publish or perish" is usually considered applicable only to academia. The college-level engineering teacher is expected to do research and publish the results. The consulting or industrial engineer or technician can also gain credibility, however, if he or she can effectively write reports and give oral presentations.

As is true of most engineering knowledge, the fundamentals of report writing may be expressed in simple terms, although considerable exposition may be necessary to make them clear. In this book, I will attempt to follow these rules in order to make the fundamentals of report writing clear and understandable.

You will notice that most of the examples and the general discussions are influenced by my background in the field of heating, ventilating, and air conditioning. The principles discussed here, however, are applicable to all engineering and technical disciplines.

As you read this book, you will find frequent references to *credibility*. The term means that the client and your report's audience will be prejudiced for or against you in many ways. In general, you will begin with a degree of credibility: you are assumed capable of producing the report. You can improve or weaken that opinion, sometimes drastically, by the appearance and organization of the report. These factors, in the end, are crucial to the acceptance or rejection of the recommendations you make; in other words, they affect the value of your report.

1

Fundamentals

The purpose of a report is to inform. An engineering report describes the definition and logical analysis of a problem, including alternative solutions and, usually, recommendations.

The purpose of a report, therefore, is to define problems and solutions, to evaluate and compare those solutions, to recommend the solution(s) that the report writer believes to be best, and to provide the client with a rational basis for evaluating those recommendations. The report is an exercise in research, logical analysis, and communication. If any of these elements are lacking or poorly done, it is not a good report and will probably fail in its purpose (TABLE 1-1).

Many studies are made and reports are written because a system does not work as expected by the client. In this situation, a good engineering report takes a positive approach and looks for solutions rather than trying to fix blame.

A technical paper describes the results of applying a specific solution or solutions to a problem. In this text, I will address primarily the engineering report. It should be noted, however, that a technical paper and an engineering report have a great many elements in common. These common elements include the problem statement, background information, alternatives, and descriptions

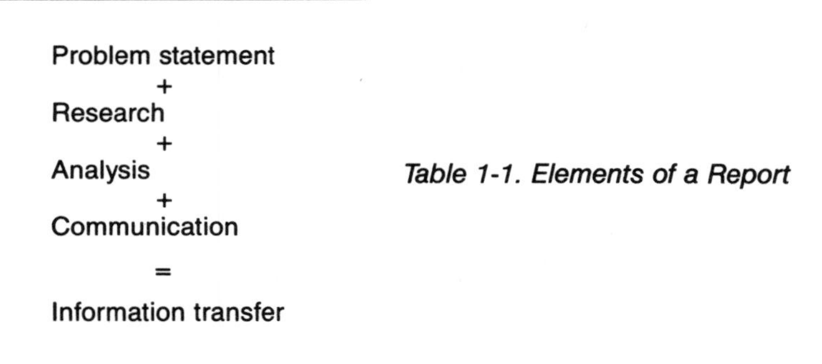

Table 1-1. Elements of a Report

of procedures. The reports differ in that the engineering report recommends implementation of an alternative solution, while the technical paper describes the results of an investigation or test. Technical papers often form part of the database for a study and report.

THE AUDIENCE FOR THE REPORT

The audience for the report consists of two groups with diverse interests. The first is the management group. These people are interested primarily in the financial implications—what will it cost if the recommendations are implemented? What will the energy savings be, if any? There may also be some interest in the problems and costs of staffing and training.

The second is the operating and maintenance group. This group is interested in the problems of staffing and training, as well as the energy cost. Their primary interest, however, is in how well the proposal satisfies the technical requirements. They want the system to work.

TYPES OF REPORTS

A report may be as simple as a one-page letter or as elaborate as a lengthy, formal dissertation that includes graphics, references, and a customized cover.

Broadly, reports fall into four classifications: letter, formal and informal, technical paper, and oral presentation. The difference between formal and informal is subjective and largely a matter of style and organization rather than content. In this text, no distinction will be made between the two.

With any written report, the complete text with graphics is delivered to the client. An oral presentation may also be made. With a report intended for oral presentation only, it is desirable to provide a set of the figures to the listeners so that they may make notes. Technical papers are intended for publication or for presentation to a meeting of a technical society before the paper is published.

ELEMENTS OF A REPORT

The essential elements in any report are these: reason for the report, scope of work (work statement), background data and information, problem statements, alternative solutions, and recommendations. I will consider these briefly here and discuss them in detail in the following chapters.

The reason, basis, or justification for a report is always the need to solve a problem. The usual term here is *authorization.* Some person or organization authorizes the investigator to study a problem or problems and to recommend solutions. In the normal client-consultant arrangement, the authorization is usually a contract that defines a scope of work, a fee to be paid, and a schedule.

Careful definition of the *scope of work* is essential. A clear definition that is contractually fair to both parties is sometimes difficult to make. This topic is discussed in chapter 3.

Data acquisition includes a field investigation of existing conditions, review of design documents for proposed or existing facilities, and a study of both the manufacturer's literature and the general literature on these types of problems, calculations, and analysis. Data acquisition and presentation are discussed in chapters 4 and 5.

The *problems* must be clearly stated so that solutions may be shown to be applicable. Much of the time, you will find that the problems outlined in the scope of work are incomplete, inaccurate, or poorly defined. A field investigation should result in better problem definition and often will uncover other, related problems. Typically, the client is aware that there is a problem but has no clear idea of what it is. Thus, the scope of work often requires the consultant to define the problem as well as the solution. For a full discussion of problem definition and solution see chapter 3.

Alternative solutions are developed out of the problem definition and the database information. A report writer needs to consider both traditional and nontraditional solutions, though always within the bounds of good engineering practice. The simplest solution is usually the best; complex systems are difficult to operate and maintain. Typically, the rule to remember is that the system operator will reduce or simplify the system to his level of understanding. (See chapter 6.)

Criteria for evaluating alternatives and making *recommendations* include simplicity, reliability, controllability, ease of installation, and cost. The client's desires must also be considered, together with code requirements. Intangibles, such as experience and bias, may often influence the final recommendation. This topic is discussed further in chapter 6.

LOGIC

Throughout this book you will find frequent references to *logic*, *logical development*, and the like. All technical writing must use the rules of logic in presenting information. Without a logical development of the thesis, little or no information will be conveyed. A brief discussion of this topic is therefore needed.

Most logic is based on a *syllogism*. A syllogism is a series of three statements. The first two are called the *major premise and minor premise*, the third is the *conclusion*. The conclusion follows as a fact if the premises are accepted as facts. A simple illustration is:

a. Evergreen trees never shed their leaves
(major premise)
b. Pine trees never shed their leaves
(minor premise)
Therefore,
c. Pine trees belong to the evergreen family
(conclusion).

A *logical chain* is a series of syllogisms in which each conclusion becomes a premise of the next syllogism. Thus readers are led to a final conclusion, which can be accepted as fact only if the chain is complete and each premise in the chain has been proven or explained. Each argument, whether technical or philosophical, must follow this logical line of reasoning, development, and proof if it is to be accepted. Many of the steps may be simple explanations, which readers will accept as reasonable or correct at face value or because they accept the expertise of the presenter. New or revised data at any step may cause them to reconsider their conclusions. That has happened to me many times.

Another type of logic is the *action-reaction* format: IF A THEN B. This is readily seen in an electrical circuit: IF the switch is closed THEN the light comes on. Forms of this type, which are used in computer logic, include AND, OR, and NOR statements. The AND statement is analogous to the syllogism, IF A AND B THEN C. The OR statement is logically different but similar: IF A OR B THEN C. The NOR statement is negative: IF NOT A NOR B THEN C. There is also a negative AND statement, called a NAND. These statements, when combined with statements for "equal to," "greater than," and "less than," provide a complete computer logic system.

When a technical writer fails to inform or communicate properly, the problem is often due to a missing link, or links, in the logic chain. You must build your logical structure brick by brick, starting with a foundation of fundamentals. Even though the audience is very sophisticated, you cannot assume that some basic fact is understood by all; you must expound and define it. If, in the process, you bore some of the audience that's too bad. But you can't talk just to a favored few—perhaps even those few will not know all the things you expect of them. Always assume, then, that everyone in the audience needs to understand each step of the logic and might need a reminder or clarification even when they "already know that!"

The difficulty is to avoid oversimplification. Consider your audience and their general level of education. Then let that level influence the style of your writing. For a report, the audience may include people with many different levels of technical understanding. For a technical paper that will be presented to a technical society, you can assume a higher level of understanding. But, by all means, avoid the kind of writing that strives to impress, rather than communicate! Multisyllable words are not a substitute for logic.

THE VALUE OF A REPORT

Studies are made and reports are written for many reasons, all of them basically economic. The client needs to make an informed decision. The report must give him the information he needs to determine both economic and engineering feasibility. The people who are concerned with financing the proposed solutions need justification that the money will be well spent. As a report writer, you must also be concerned with justifying your fee and maintaining your credibility. Therefore, you must do a thorough job of researching and analyzing to be sure that your recommendations are objective, factual, and solve the problems of the client. To the extent that the report accomplishes these aims, it has value for everyone.

SUMMARY

A technical report must have a purpose, be directed to a specific audience, be well researched, and have a logical organization and exposition in order to be of value.

2

Language and Communication

The *New World Dictionary* defines *meaning* as: What is intended to be, or in fact is, signified, indicated, referred to, or understood. The words and graphics used in a report may have a specific meaning for report writers, but may have a different meaning for the reader. When this occurs, communication is garbled and the information transfer is flawed.

It is necessary to choose words carefully and to provide definitions when there is doubt. This is especially true for technical terms. The audience for the report is not always technically trained. Managers, financial advisors, and even other engineers may not be familiar with the exact meaning of new or exotic words and phrases. The best rule is: when in doubt, define.

Often, the context in which a word is used may help clarify its meaning. Context may also obscure meaning, however. The relationships among words are important.

Another, very important factor is the bias of the reader. All of us tend to filter information through our preconceived ideas. To a great extent, we hear and read what we *expect* to hear and read. Only careful, factual writing has a chance to cut through prejudices. The elements of communication (TABLE 2-1) are defined below.

Table 2-1. Elements of Communication.

1. Grammar and syntax
2. Style
3. Definition of terms
4. Objective writing
5. Logical structure
6. Bias of audience

THE MECHANICS OF WRITING

Despite the debasing of the English language by careless writers of every form, the rules of English grammar are still in force. If a report is to be credible, it must be grammatically accurate and have correct spelling and punctuation. The pitfalls of syntax, case, and agreement are many and varied. While this text is not intended to teach grammar, a few examples may be helpful.

Syntax refers to the structure of a sentence. Every complete sentence must contain a subject and a verb. Other parts of speech—adjectives, adverbs, prepositions, conjunctions, interjections—may also be present and must be used correctly when they appear. You may find incomplete sentences in fiction, essay, and other forms of emotive writing, but emotive words and phrases have no place in a technical report.

Agreement refers to the relationship between subject and verb: singular form to singular form, plural to plural. Part of the problem is recognizing the form of the subject. A common example, since the beginning of the computer age, is the word *data*. This is a plural noun. The singular is *datum*. A great many people treat the word data as though it is singular and say "data is...." Another common source of error is the compound subject—"A and B *are* different" and "A or B *is* acceptable"—both of which are correct as shown.

Case refers to the subjective, objective, and possessive forms of some nouns and pronouns. Examples of correct usage are: you and I will go (subjective), versus they sent for you and me (objective).

There is no excuse for a misspelled word, though "typo!" is always a good alibi. As writers, we must keep a dictionary handy. Many word processors now include automatic spelling checks. A useful tool is *The Bad Speller's Dictionary*, which is published by Random House.

Punctuation is a problem for most of us. There are rules regarding the use of commas and semicolons. These punctuation marks set off subordinate phrases and clauses and separate items in a series. Sometimes punctuation is used to obtain emphasis. Sometimes the rules may be bent or broken for the sake of clarity. The overuse of commas is more common than the underuse of these marks. Be sure that each comma you use clarifies the phrase in which it is used.

STYLE

Writing style is the result of the writer's use of mechanics, as described above, and many other things. There are numerous authorities on style, none of which are final or absolute. Many magazines and newspapers provide style books for their writers. These include instructions on punctuation and grammar. They may also include forbidden words and phrases that are not to be used because they are considered offensive or controversial. In technical writing, you must observe the general rules of grammar, punctuation, spelling, agreement, and case.

Rules for handling numbers, abbreviations, citations, and references are published by many technical societies. Numbers in the text should be written out if the value is nine or less. For larger numbers, figures may be used. Avoid the use of a number at the beginning of a sentence; it may be mistaken for a paragraph number. The American Society of Heating, Refrigeration, and Air Conditioning Engineers (ASHRAE) publishes *Instructions for Authors*. It also lists many approved abbreviations and terms in the *ASHRAE Handbooks*. The American Standards Association (ASA) publishes a lengthy list of abbreviations, many of which are republished in other societies' journals. The general rule is that periods are not used with abbreviations except to avoid confusion, i.e., "in." for inches. In other chapters and examples in this book, I will describe my preferences in some of these areas.

Style also includes sentence structure and organization. Some writers seem to feel that they are being paid by the word or that the thickest report will be given the most credence. The clearest writing, however, is also the most succinct and least verbose. Use ordinary words to increase understanding. Dr. Richard Wydick has written an excellent piece titled "Plain English for Lawyers."[1] In this article, he expands at some length about lawyers' habits of using excess verbiage. Some examples: "because" becomes "for the reason that;" "on" is sometimes expanded to "with respect to;" and "at that point in time" simply means "then." There are many such examples. Most first draft writing can be reduced from 20 to 50 percent by judicious attention to the removal of excess words.

Another common failing of the report writer is the introduction of a dependent clause or phrase between the subject and verb. By the time the reader arrives at the verb, he has lost track of the subject. Agreement between subject and verb may also be lost in the fog of words. An example of poor usage is: The control system, which was custom-designed using a mixture of electronic and pneumatic components, was very complex. It would be clearer to simply state: The control system was complex. Then use additional sentences to describe why it was complex.

Pronoun antecedents must be clear. The alternative is to avoid the use of pronouns. This type of error always leads to confusion and misunderstanding.

Slang expressions have no place in technical writing. I recently observed such statements as "whopping amount of energy" and "the owner had a lot at stake" in an entry in an engineering competition. This, to me, was detrimental to the entry.

A good rule to remember is that short sentences increase comprehension. The extreme simplicity of "See Spot. See Spot Run." is not a style that should be cultivated either. Variety is the key to maintaining interest. As Mark Twain put it, "At times (the writer) may indulge himself with a long (sentence), but he will make sure there are no folds in it, no vaguenesses, no parenthetical interruptions of its view as a whole; when he has done with it, it won't be a sea-serpent with half its arches under the water, it will be a torch-light procession."[2]

OBJECTIVE VERSUS SUBJECTIVE WRITING

The ideal for which every technical writer should strive is that of objective writing. That there is no such thing is because of the built-in bias of every writer, which is the result of his experience and education. Each writer must try to understand the extent of his biases and allow for them in the analysis and presentation of the facts. Sometimes the subjective approach is the only way to make a difficult decision, such as when facts are missing or unclear: There isn't any reason for it, that's just the way I do it! That kind of decision-making must be based on long experience and proven expertise. It should be used only as a last resort and even then may be wrong.

What distinguishes objective from subjective writing? Objective writing presents facts, which are determined from research, observation, and calculation, and presented in a logical, orderly manner. Subjective writing is short on facts but long on conclusions. It is full of adverbs, adjectives, and emotive words and attempts to convince in spite of the evidence. Subjective writing has no place in a technical report, but may creep in when the writer has a point of view he wants to maintain. The truly professional writer will watch for this and avoid it. The reader may not always note subjective writing, but he will be aware, in a vague way, that something is not quite right.

Subjective and objective writing can be illustrated by other terms:

- *Facts* are data that can be observed or deduced from observation and generally mean the same thing to any knowledgeable person. Such facts are objective. Unfortunately, not all facts are clear and agreed upon so even here some subjectivity may be present.

- *Opinions* and *judgments* are, or should be, based on understood facts. Opinions may or may not be valid, depending on the accuracy of the facts and the experience and training of the researcher. Two experts who are given valid data should arrive at similar opinions. The accuracy of the database should be the governing factor. Opinions may be used in a report, but should be clearly defined as such.

- *Hypotheses* and *theories* are the basis for research. An hypothesis is an idea. When it appears that the idea may have merit it becomes a theory, which may be tested in various ways to determine its validity. A proven theory becomes a fact and part of the database.

- *Generalizations*, *guesses*, and *conclusions* are based on an accumulation of circumstantial evidence. In the absence of factual data, generalizations may be useful but may not be used as facts. A generalization is sometimes called an educated guess, since some degree of expertise must be applied.

- *Emotive words and phrases* have been mentioned before. A technical report is a tool for selling the writer's opinions, but the presentation must be factual, describe alternatives, and avoid emotional pleas. Intangible and emotional reasons, such as the client's preferences, may enter into the final recommendations. These must be addressed as what they are, however, and not disguised as facts.

THE POWER OF THE POSITIVE

A widely read book today is *The Power of Positive Thinking*, by Norman Vincent Peale. In his book, Dr. Peale talks about the use of positive thinking to overcome adversity as contrasted with negative thinking, which succumbs to adversity. This philosophy is an appropriate approach to technical report writing. As the writer studies a problem, it is easy to see that it will be difficult and expensive to provide a solution. A solution may not even be possible. The whole point of the engineering approach, however, is to find a solution, so the writer must think positively. The only solution may be expensive and difficult, but that is not an excuse to give up. Writers must always look for the positive alternative, even if it goes beyond the original scope. When this approach is taken, it is often surprising what positive things happen.

SUMMARY

The meaning of the words in a report must be made clear by using context and definition. Reports must be as objective as possible. When an opinion or generalization is used it must be noted as such.

REFERENCES

1. Richard C. Wydick, "Plain English for Lawyers," *The California Law Review*, 66, no. 4 (July 1978).
2. As quoted in E. Gowers, *The Complete Plain Words*, Fraser rev. ed. (Boston, MA: David R. Godine Publishers, Inc., 1973), 183.

3

Organization of a Report

Since a report is an exercise in logical exposition, it follows that the elements of a report should be arranged in a logical progression. This chapter will discuss one method of arrangement that has worked very well for many years. It is based on the assumption that there are two different audiences for each report. One is the *technical* audience: those who understand and are interested in the details, calculations, and analysis. The second is the *management* audience: those who are interested in the alternatives, recommendations, and costs. The management audience depends on the technical group to check the validity of the technical details.

A report that serves both audiences must begin with a succinct but comprehensive summary, which is sometimes called an *executive summary*. This summary should be followed by a lengthy discussion of the details on which the summary is based. The actual writing process should begin with the details. The summary should be written last.

ELEMENTS OF A REPORT

A typical report outline is shown in TABLE 3-1. The details may vary from one report to another, but the basic elements invariably should be present. The summary and

Table 3-1. A Typical Report Outline.

I. SUMMARY AND RECOMMENDATIONS
   A. Authorization and Scope
   B. Acknowledgments
   C. Summary
      1. Existing Conditions
      2. Problems
      3. Alternative Solutions
   D. Recommendations
II. EXISTING CONDITIONS AND PROBLEMS
   A. Existing Conditions
   B. Database
   C. Problem Statements
III. ALTERNATIVE SOLUTIONS
   A. General Criteria
   B. Database
   C. Alternatives
   D. Cost Analyses
IV. APPENDIX
   A. Glossary of Symbols and Abbreviations
   B. References
   C. Calculations
   D. Other

recommendations in the first section are a short restatement of the material in the other sections. In most reports, there will also be figures and tables. Summary tables are especially useful if properly handled (see chapters 5 and 6). All of these elements will be discussed in detail.

FORMAL AND INFORMAL REPORTS

The report arrangement may be more or less formal, depending on its type. A *letter* report (FIG. 3-1) may be used to address a single, uncomplicated problem. The letter report must contain all of the basic elements, but should not have headings and subheadings. It contains essentially the same material as the summary and recommendations section of TABLE 3-1, but is written in an informal style. It may include one or two simple figures or tables and some calculations. A letter report is often used to provide a quick response to an inquiry. The report writer, however, must not let the need for haste override the need for careful and thorough analysis of the problem.

A *formal* report should be arranged in outline form. It should have headings and subheadings, a title page as in FIG. 3-2, and a table of contents as in FIG. 3-3. This type of report may deal with only a few simple problems or it may be as detailed and extensive as a complex conceptual design or master plan. More often, its purpose falls somewhere between the two extremes. In every case, the basic elements of communication and logical development must be present.

DEFINING THE SCOPE OF WORK

Perhaps the most difficult part of any study/report is that of defining the *scope of work*, sometimes referred to as the *work statement*. The definition must be so clear that when the report is presented, the writer can say, "This is complete in accordance with the scope of work," and have the client agree. This kind of clarity requires your best writing effort. As has already been noted, the words you write will probably not have the same meaning for all of your readers. Since the consultant's fee must be based on the scope of work, any serious misunderstanding regarding its meaning can lead to loss of confidence on the part of the client and possible financial loss for the consultant, who may find himself doing more work than he expected.

The scope sometimes begins with a work statement from the client. The consultant should be sure that he understands the meaning of the work statement before proposing a fee. This will usually require a discussion with the client. In many cases, the client lacks a clear understanding of the problem. Thus, the scope may include problem definition or the problem may be defined in preliminary discussions. In either case, the scope definition may be somewhat vague. Most scope definitions are developed during discussions between

Roger W. Haines, P.E.

November 14, 1989

Mr. John Doe
Richard Roe Company
P.O. Box 0001
Anywhere, CA 90000

Dear Mr. Doe:

In our phone conversation of November 3 you asked me to give you an analysis of the available devices for sensing and controlling relative humidity (RH) in a space. Your specific application is in process control where the specifications call for controlling RH within plus or minus 5% (absolute) or even closer.

We need to consider not only the initial accuracy but also the loss of accuracy over time due to hysteresis and drift, as well as the maintenance required to retain accuracy or to recalibrate. Calibration of humidity sensors is not easy--the most serious problem is that of determining the real value of RH in a duct or space.

Sensors using dimensional change materials are not suitable. There is simply too much hysteresis and drift. The older hygroscopic sensors were accurate but required a great deal of maintenance to retain that accuracy. The best devices we have at present (and can recommend) are:

1. Solid state sensors, using deposition techniques to provide a thin or thick film of hygroscopic material. Either resistance or capacitance varies as a function of RH. The best of these devices can be specified to have an absolute accuracy of plus or minus 3% with a drift not to exceed 1% per year. They can be obtained for duct or space mounting.

2. The "Chilled Mirror" sensor. The name describes the operation. A light source is "read" by a photocell. The light is reflected to the photocell by a stainless steel mirror which is provided with a thermoelectric cooling system. The dew point is sensed by noting the change in reflected light when condensation occurs on the mirror and reading the mirror surface temperature. This can be converted to RH if desired. The device is very accurate and can be obtained for duct or space mounting. Some models can be used for calibration of other humidity devices. Maintenance consists of occasionally cleaning the mirror. There is no loss of accuracy over time.

Regardless of the sensor selection, the controller must operate in the PI (proportional plus integral) modulating mode. "Proportional only" is not suitable due to the inherent offset (error) of this mode, which can exceed the possible sensor error.

It is recommended that electronic control devices be used, since the sensors described above are available only in electronic form. We have used both of the recommended sensor types successfully in several installations.

If there are any questions please call me.

Sincerely

Roger W. Haines

Fig. 3-1. An example of a letter report.

REPORT ON
HVAC CONTROL SYSTEMS
at
STATION
for
PROJECT
DEPARTMENT OF
CITY OF

by

Roger W. Haines, P.E.
Consulting Engineer
Laguna Hills, California
September, 1988

Fig. 3-2. An example of a title page for a report.

the client and consultant in the hope that both will understand the criteria in the same way, though this doesn't always happen.

The scope of work or work statement must include:

- A definition of the problem or problems to be solved. If part of the work is to define the problem(s), then a definition of the area in which the problem(s) falls must be included.
- A definition of the depth of the study that is to be made. The study may simply be a definition of the problem or it may include alternative solutions and recommendations. It may include detailed or rough cost analyses or no costs. It may include conceptual designs or detailed design data, including calculations and outline specifications.

TABLE OF CONTENTS

SECTION	TITLE	PAGE
I	SUMMARY AND RECOMMENDATIONS	I-1
	A. Authorization and Scope	I-1
	B. Acknowledgments	I-1
	C. Summary	I-4
	D. Recommendations	I-4
	Table I-1 Summary of Problems, Alternative Solutions, and Recommendations	
II	EXISTING CONDITIONS AND PROBLEM DEFINITION	
	A. Existing Conditions	II-1
	B. HVAC Control Problems	II-4
	Fig. II-1 Controls and Service Buildings—Floor Plan	
	Fig. II-2 Halls and Auxiliary Buildings—Floor Plan	
	Fig. II-3 ASU-2 Air Handling Units—Halls	
	Fig. II-4 ASU-1 Air Handling Unit—Halls	
	Fig. II-5 Cooling Water System—Schematic Diagram	
	Fig. II-6 ASU-4 Air Handling Unit—Control Building	
	Fig. II-7 ASU-5 Air Handling Unit—Service Building	
	Fig. II-8 Portion of Existing Pneumatic Control Diagram	
	Fig. II-9 Air Handling System for Cooling Tower Building	
	Fig. II-10 Wind Eddies at Cooling Towers	
III	ALTERNATIVE SOLUTIONS	
	A. General	III-1
	B. Alternative Solutions	III-1
	C. Cost Analysis	III-5
	Fig. III-1 Electric/Electronic Controls for Hall Building—six sheets	
	Fig. III-2 Electric/Electronic Controls for Control/Service Buildings—four sheets	
	Fig. III-3 Psychometric Chart for Evaporative Cooling—Single- and Two-Stage	
	Fig. III-4 Cooling Tower Stack Extensions	
IV	RECOMMENDATIONS	
	A. General	IV-1
	B. Recommendations	IV-1
	Table IV-1 Summary of Recommendations	
V	APPENDIX	
	A. Control Theory	V-1
	B. References	V-3
	C. Outline Specification for Electronic Controls	V-3
	D. Symbols and Abbreviations	V-3
	Fig. V-1 Elementary Control Loop	
	Fig. V-2 Proportional Control Action—Stable	
	Fig. V-3 Proportional Control Action—Unstable	
	Fig. V-4 Proportional plus Integral Control Action—Unstable	

Fig. 3-3. An example of a typical table of contents.

- Client criteria as to budgets, time schedules for both the report and subsequent physical modifications, availability of client's people for assistance and information to the report writer, availability of documentation, access to client's premises, and any special criteria.

The consultant's fee is based on the scope of work and may be proposed in one of three ways:

- A lump sum that covers all charges.
- An hourly rate plus expenses with a maximum charge that may not be exceeded. Rates also may be established for various classes of work: engineering, drafting, clerical, etc.
- An hourly rate with no limit. This may be used when the problem is poorly defined and the difficulties of definition are unclear. This type of fee may be used when there is a lack of documentation and the amount of time and effort required by the consultant to obtain the data cannot be reasonably estimated.

THE PROBLEM-SOLVING PROCESS

Problem solving for the engineer/technician is a logical process. The process should proceed through definition, data acquisition, and analysis to alternative and recommended solutions. An outline that writers have found useful is shown in TABLE 3-2 and discussed below.

Define the objective. What is it that you are really trying to accomplish? A clear problem definition can come only from a clearly defined objective. For example, the objective may be to provide an appropriate environment for a specific process. Problems that could arise from this objective include definitions of the environmental criteria and selection of the types of heating, ventilating, and air conditioning (HVAC) systems and controls that would be required to satisfy those criteria. The

Table 3-2. The Problem-Solving Process.

1. Define the objective.
2. Define the problem.
3. Collect data.
4. Define alternative solutions.
5. Evaluate alternatives.
6. Check.
7. Select an alternative.
8. Implement the selected alternative.
9. Evaluate.

environmental objective could also include such things as noise, lighting, and space for equipment.

Define the problem. As stated above, the problem definition develops from the objective. In an existing system, it might be that the environmental criteria are not being met and the problem is to determine the modifications necessary to meet those criteria. Harry Lorayne put it very succinctly, "Most problems precisely defined are already partially solved."[1]

Collect data. Data sources include documents, studies of existing conditions, interviews with the client's operating and engineering personnel, calculations (including computer analyses), equipment information from manufacturers, criteria for system accuracy and reliability, client criteria for budgets and schedules, etc. Among the best data sources are the people who use, operate, and maintain the systems. These data must be filtered through your experience since they tend to be one-dimensional because the person responsible for a particular piece of information may not see the complete picture.

Define alternative solutions. There are always several ways to solve any problem. One alternative is to do nothing. While this has no first cost, it may have a high cost in other ways. It is an alternative that must be considered. A common trap is the safety of the familiar. Familiar and proven methods should not be neglected and may often be best, but you should still consider new and innovative solutions. The opposite side of the coin is that writers sometimes get overenthusiastic about new ideas and neglect to make a careful analysis and evaluation. Definition of alternatives may also include costs. Technical aspects usually should be the most important factors.

Evaluate alternatives. Evaluation includes technical and cost factors. It should also include client criteria for budgets and scheduling. In many cases, it is necessary to schedule modifications and replacements that avoid interference with the client's operations. Your evaluation must be as objective as possible, with due allowance for the subjective factors inherent in your experience.

Make an objective check. Check to see that the proposed alternatives really address the problem. It is essential that this checking step be objective. Will your proposed solution solve the problem or is it cosmetic—something you think the client will like, but that doesn't address the source of the problem? A typical example of the latter is the addition of a computer system to solve a control problem. Since the computer is simply a more sophisticated controller, it cannot solve a fundamental control or system problem and may even make it worse.

Select an alternative. This step develops from the evaluation step. In the report, you should provide a detailed recommendation of the selected alternative, together with the reasons why you think it is best. Sometimes the client accepts your recommendation, sometimes he does not. That is his prerogative. You should not allow what you think the client will like to influence your recommendation.

Implement the selected alternative. Implementation happens as a result of the report and is not really a part of this discussion. It is important, however, because only after something is physically accomplished can you determine the accuracy and value of your report.

Evaluate. Once the system is constructed or modified as recommended in the report, you can determine how good your analysis was and learn how to do it better the next time. This is called learning by mistake and is the reason writers need to study. Seminars, books, technical papers, etc., allow writers to learn from the mistakes of others.

The problem-solving process is implicit in the report arrangement shown in TABLE 3-1. The objective and problem statements come in the scope of work. Section II of the report provides for data acquisition and analysis, alternatives, and recommendations. Implementation and evaluation are not part of the report, but usually result from it.

DEALING WITH MATHEMATICS

Many technical reports include mathematical symbols, equations, and calculations. How these are expressed in the report becomes, to some extent, a function of the sophistication of the typing system being used. Most word processors and typing systems—there are very few simple typewriters any more—have the capability of setting up complex equations. It is much easier with some systems than with others. Not all systems include a complete set of mathematical symbols such as integral and infinity signs and Greek letters. Writers must suit their mathematical exposition to the equipment. For example, the equation for calculating cubic feet per minute (CFM) in an air handling system is usually written:

$$\text{CFM} = \frac{Q}{\text{AF} \times \Delta\, T.}$$

This arrangement is difficult and time consuming on many word processors. It may be simpler and just as effective to write it like this:

$$\text{CFM} = Q\ /\ (\text{AF} \times \text{Delta T}).$$

When the square root symbol ($\sqrt{\ }$) is not available, a superscript $^{(1/2)}$ means the same thing. Other symbols, such as the integral sign, may even be drawn in manually.

All terms used in equations must be defined. Definitions must include meaning and engineering units.

As a general rule, mathematics should be kept to a minimum in the body of the report. Your readers are really interested in what the results are and what they mean. Most readers bog down in mathematical detail. For those who wish to check the math or learn from it, detailed calculations may be included in the appendix. Some mathematics may be necessary in the summary or elsewhere to clarify the discussion. Common sense and experience should be used to evaluate the need for equations in these sections.

FORMAT AND APPEARANCE

The appearance of your report can have a great influence on its credibility or lack of it. It takes just a little extra effort to provide a neat, pleasing package for your ideas, and it makes a big difference. Format, typing, binding, and reproduction methods all affect the appearance.

Format. Format includes an orderly and logical arrangement as discussed above; the use of a standard form similar to that in FIG. 3-4, which lends uniformity; neatness—no strikeovers or gaps showing last-minute deletions; and the like. As already emphasized, accurate grammar, syntax, and spelling also are required.

Typing. Most offices today have word processors or intelligent electric typewriters or both. Ideally, software packages that allow proportional spacing and right justification provide the best appearance (FIGS. 3-5a and 3-5b). This type of software is not expensive, is readily available, and is easy to use. It is available for both word processors and intelligent typewriters. Most of these software packages also provide features such as automatic page numbering, headers and footers, and the elimination of widow lines, which are single lines that end or begin a paragraph at the top or bottom of a page. The printer should be a daisy-wheel or laser type. Dot matrix printing has been improved, but in my opinion still does not equal the other methods (see chapter 7).

ROGER W. HAINES * CONSULTING ENGINEER * LAGUNA HILLS, CALIFORNIA

Fig. 3-4. An example of a standard report form. The form has been reduced for publication. The actual size is 8 1/2 × 11 inches.

This paragraph is written to illustrate the difference in typing formats. It is typed using a 10-point type face with uniform spacing between letters, six lines per inch and without right justification. (It has been reduced for publication). It looks like a typed page rather than a printed one.

Fig. 3-5a. An example of a paragraph typed with uniform spacing and without right justification.

Reproduction. Photocopying is probably the most widely used method of reproduction. If the copier is well maintained and each copy is inspected for flaws, this method works quite well. In normal practice, the copier does not make consistently perfect copies and writers have to settle for something less. If you follow the recommendations on typing in the previous paragraph, you will have camera-ready copy that can be reproduced by photo offset printing. This method of reproduction provides a good, consistent quality and should be used for the best appearance. With this method, you can easily print on both sides of the paper and reduce the volume of long reports. On short reports that are 20 to 30 pages, it is preferable to use only one side of the paper.

Binding. There are numerous binding systems available. For short reports, there are a variety of clamping type binders that provide a good appearance and are easy to assemble. For larger reports, punched hole or spiral binding methods may be used. For a really professional appearance, preprinted standard covers may be used. These covers can be designed in any format suitable to the consultant's background and specialties. They should be printed on heavy weight paper, usually

This paragraph is written to illustrate the difference in typing formats. It is typed using a proportional type face with proportional spacing between letters, six lines per inch, and with right justification. (It has been reduced for publication). It looks very much like a printed page. It also uses fewer lines than uniformly spaced type for the same amount of text.

Fig. 3-5b. An example of a paragraph typed with proportional spacing and right justification.

40- or 50-pound stock. A standard cover may be overprinted for each specific project.

REFERENCES AND CITATIONS

All references and citations should be acknowledged in the proper format, which is similar to the format used for the references in this book. This is good form and allows the reader to study the reference if he so desires. References may be placed at the end of each section of the report or grouped in the appendix.

REFERENCES

1. Harry Lorayne, ''Memory Makes Money,'' quoted in the *Reader's Digest*, (November 1988), 33.

4

Data Acquisition

The term *data* in this chapter refers to the information you accumulated when you studied the problems defined in the scope of work. Data include objectives, studies of existing conditions, applicable codes and standards, client criteria, calculations, manufacturer's catalogs, cost information, the advice of friends, colleagues, and paid consultants, and your own standards, experience, and preconceptions.

The importance of good data should be obvious, but cannot be overemphasized. The reliability of the report is directly related to the reliability of the data. It is only natural that you evaluate all data in terms of your experience. If your personal experience is insufficient, then you must rely on the advice of others whose experience you trust. No one can be a competent expert in all the aspects of a complex problem. There is no stigma attached to the use of other experts when needed. In fact, such use will probably increase the credibility of the report.

DOCUMENT SEARCH

The first step in data acquisition is a document search. There is a considerable difference between a search that relates to a facility that is still in the design stage and one that relates to an existing facility.

Proposed facilities. In this case, the documents consist of a set of criteria and, perhaps, some incomplete drawings and specifications. If the investigation and report have been scheduled at the proper time in the design procedure, there will be some flexibility in the available alternatives. There will always be some limiting factors, particularly as to space and budget. Hopefully, you will be given the opportunity to have some input into the design process. In the best circumstances, you will get the space and budget you need to satisfy the criteria while allowing the system to be operated and maintained efficiently.

Existing facilities. The first rule regarding documentation of existing facilities is that documents probably don't exist. The second rule is that if documents do exist, they are probably wrong. So the procedure is to get any documents that are available and then make a field check to determine the accuracy of the documents. The operation and maintenance people are your best resource. They know what is where and what they have to do to respond to complaints. In larger facilities with a plant engineering group, you may get better documents and more help, but even here documentation should be suspect.

FIELD STUDIES

The purpose of a field study is to determine, as precisely as possible, existing conditions. This involves much more than simply looking at equipment. TABLE 4-1 lists some of the general areas to be researched. Specific problems may indicate the need to research other areas.

Review existing conditions. Compare existing conditions to available documents and revise those documents to show as-built conditions. As already noted, few if any existing documents are accurate. Before you can solve any problem, you will need to understand the existing conditions since they will be the starting point

Table 4-1. Field Study Checklist.

1. Compare existing conditions to available documents; revise documents to match.
2. Where documents do not exist, provide as-built documents based on field check of existing conditions.
3. Interview operating, maintenance, and engineering personnel for information on history, operating procedures, training programs (if any), and suggestions.
4. Interview people who live and work in the facility.
5. Take measurements of and observe operating conditions.
6. Note space and access availability for modifications.
7. Note condition of existing equipment, age, quality of maintenance, and suitability for the service.

for any modifications. Where documents do not exist, it will be necessary to make new documents. For study purposes, these documents need be no more than simple sketches or notes. These sketches and notes can be used, as needed, to develop graphics for the report.

Interview people. Talk to the people who maintain and operate the facility. Talk to the engineering staff, if that is a different group than operations. Talk to the people who use the facility. Ask them how they perceive it should work. Interviews don't have to be formal and probably work better in an informal one-on-one setting. Most people have opinions on the systems you are studying and welcome the opportunity to extend those opinions, especially to someone who might be able to improve conditions and implement their ideas. You will find that you may receive a mass of conflicting and contradictory data, but among all the inapplicable information will be a great deal that is helpful. Some data may even be pure gold. The secret to successful interviewing is to listen with no preconceptions and no condescension. Don't talk down to people. If you start by telling them what needs to be done, you will get no information. Some of the interview data may conflict with the actual conditions determined in step one because not everyone understands the situation in the same way.

Take measurements and observe operating conditions. Use only high-quality, recently calibrated instruments. Do not depend on in-place gages, thermometers, etc. Sometimes things don't operate the way the instructions say they will. The manufacturer's data may be inaccurate or field conditions may prevent equipment from working as it does during tests in the laboratory. Proportional controllers never control at the set point—there is always an offset. This means that design conditions are seldom achieved, even when the equipment capacity is adequate. If non-design conditions are acceptable, this may not be a problem. What do the criteria say? Air handling unit fans almost never perform in the field as shown in the catalog simply because field conditions never match lab test conditions. The Air Moving and Conditioning Association, Inc. (ACMA) recognizes this and provides a manual that allows writers to estimate the effects of system geometry on fan performance.[1]

Note space and access availability for modifications. Many equipment rooms are far from spacious or accessible, which makes modifications difficult. This affects cost and may even make some alternatives impossible. Space modifications may be necessary, however, in order to solve the problem. In many equipment rooms that have adequate space, you may find that the space has been taken over for storage. This is usually hazardous and probably illegal. Your report will need to address this kind of situation.

Note condition of existing equipment. If equipment is old or has been poorly maintained, replacing it may be preferred to repairing and maintaining it. If the equipment is not properly designed or sized to satisfy the criteria, it will need to be upgraded or replaced. You may find that most problems with heating, ventilating, and air conditioning (HVAC) controls are caused by improperly designed control systems or by improperly sized HVAC equipment. If the addition of computer controllers to an existing system is being planned, it certainly will be necessary to provide sensors that are suitable for the high quality controllers.

MANUFACTURER'S DATA

Any technical report that deals with equipment, particularly with existing equipment, will rely on the equipment manufacturer(s) for information and performance data. Most of this information will come from the local representative, but for difficult or unusual problems the manufacturer's specialist may provide assistance. Report writers tend naturally to turn to people they respect and who have been helpful and informative in the past. These people will provide you with accurate product performance and cost data, as well as related engineering expertise. They also will frequently offer suggestions if you share the problems. You must use care not to solicit specifications that might limit competition, unless the specific product is the only satisfactory one. So long as the representative does not take undue advantage, he will continue to merit your confidence.

CODES AND STANDARDS

It is apparent that any recommendations in the report should comply with all related codes and standards. The most important of these are the local building codes, since they are legal requirements. Building codes may be enforced by any local governing agency, usually a city, county, or state. Most building codes now include requirements that facilities be designed for the minimal energy consumption consistent with use, though this does not guarantee that the facility will be operated in an efficient manner. The latter becomes an education problem. Many times the report will have to address the need for education and training of personnel who use and operate a facility.

Standards are promulgated by a number of organizations. Federal standards are developed by the U.S. government to provide a uniform basis for design, specification, and purchase by the government. They are frequently used as references on nongovernment projects.

Consensus standards are developed by industry and professional groups to provide a uniform basis for comparing, testing, and rating products. There are many such groups. A partial list is contained in TABLE 4-2. Many of these standards are written by volunteer committees. Proposed standards are distributed to interested parties for review and comment. All comments are considered and receive a response. The final draft of the standard represents a general consensus, though seldom complete agreement. Consensus standards have no legal force, except when compliance is called for by a specification. Many consensus standards are adopted by local code authorities, however, at which time they become part of the codes and are mandatory.

Table 4-2. Codes and Standards Organizations in the U.S.

ABMA	American Boiler Manufacturer's Association
AGA	American Gas Association
AHAM	Association of Home Appliance Manufacturers
AIHA	American Industrial Hygiene Association
AMCA	Air Movement and Control Association
ANSI	American National Standards Institute
ASA	Acoustical Society of America
ASCE	American Society of Civil Engineers
ASHRAE	American Society of Heating, Refrigerating, and Air Conditioning Engineers, Inc.
ASME	American Society of Mechanical Engineers
ASTM	American Society for Testing and Materials
BOCA	Building Officials and Code Administrators International, Inc.
CABO	Council of American Building Officials
CAGI	Compressed Air and Gas Institute
CSI	Construction Specifications Institute
CTI	Cooling Tower Institute
HI	Hydraulic Institute
HYDI	Hydronics Institute
ISA	Instrument Society of America
NEMA	National Electrical Manufacturer's Association
NFPA	National Fire Protection Association
SAE	Society of Automotive Engineers
SBCCI	Southern Building Code Congress International, Inc.
SMACNA	Sheet Metal and Air Conditioning Contractor's National Association
UL	Underwriter's Laboratories, Inc.

PUBLICATIONS

This term refers to all of the written reference material used to compile your report. You may read and study articles, reference books, other reports, and anything else that is applicable in order to take advantage of the experience and expertise of others. You will learn from their mistakes. Typical materials include handbooks published by engineering societies and manufacturers, textbooks, technical articles from magazines, and previous reports on the same or related topics. Many manufacturers' handbooks are the best references available. The Marley Company book on cooling towers[2] or the Crane Company book on flow of fluids[3] are two examples.

If you use magazine articles and previous reports, it is necessary to separate theory from fact. Many such documents simply express theoretical solutions that have not been proven. Some discuss real solutions to real problems, but the problems may differ from those you face. Discernment and common sense, as usual, are needed.

CALCULATIONS

Calculations are needed for loads, equipment sizing, costs, estimates of energy consumption, sizing of transmission systems (i.e., pipes, ducts, and wires), and many other things. All calculations must be based on some generally accepted references and procedures, or on basic engineering equations. The calculations must include clear statements about reference sources, all assumptions made, equations used (with definitions of all terms in the equation), and intermediate and final results. Neglecting these things may make it difficult or impossible to reconstruct your thinking at a later date. Experience tells us that after six months most of the detail of why we did certain things has been forgotten. As previously noted, detailed calculations may be placed in the appendix, with only the results shown in the body of

DETERMINING STACK EFFECT ΔP IN XYZ BUILDING

11-15-89
RWH.
JOB. 902

REF: ASHRAE FUNDAMENTALS, 1989.
PAGE 23.4, EQUATION 9

$$\Delta P_s = C_2 d_i g (h - h_{NPL}) \left(\frac{T_i - T_o}{T_o} \right)$$

C_2 = UNIT CONVERSION FACTOR = 0.00598

d_i = air density ≈ 0.075 lb/ft^3

g = gravitational constant = 32.2 ft/sec^2

h = height of observation ft., assumed 5 ft, (median of 2nd basement)

h_{NPL} = height of neutral pressure level, ft.

T = absolute temp, °R, Assume T_o = 60+460 = 520, T_i = 530

i = inside

o = outside

ΔP_s = pressure due to stack effect, inches H_2O, ΔP at height h. —

Building height - 6 floors + 2 basements ≈ 100 ft.

Assume h_{NPL} = 60 feet for wind velocity ≈ 15 mph.

Then:

$$\Delta P_s = (0.00598)(0.075)(32.2)(-55)\left(\frac{530-520}{520}\right)$$

$$= -0.0152 \text{ inches } H_2O.$$

FOR INSIDE TEMP. of 80 F and OUTSIDE TEMP. OF 50 F

$$\Delta P_s = -0.0467 \text{ inches } H_2O,$$

ROGER W. HAINES * CONSULTING ENGINEER * LAGUNA HILLS, CALIFORNIA

Fig. 4-1. An example of a handwritten calculation.

the report. A sample calculation is shown in FIG. 4-1. Calculations may be freehand if done neatly.

OTHER CONSULTANTS AND EXPERTS

Since no one is an expert in all things, it is usually desirable and even necessary to consult with others when your own experience is not as extensive as you would wish. In most engineering offices there will be support from others, but sometimes specific knowledge may be lacking. Then it is necessary to look for other sources of help. While studying may fill the gaps, there are many areas where only experience will serve. One particular example comes to mind. Refrigerant piping is covered in several handbooks, some of which have conflicting data. There also are many things about refrigerant piping that are not in the handbooks, but exist primarily in the minds of a few old timers. This kind of resource is invaluable and sometimes hard to find (my expert is 87 years old!). The more outside sources of help you have, the better your database and, the better therefore, your report will be.

REFERENCES

1. Air Moving and Conditioning Association, Inc., "Fans and Systems," *Standard 201-AMCA Fan Application Manual* (1973).
2. The Marley Cooling Tower Company, *Cooling Tower Fundamentals*, 2nd ed.
3. The Crane Co., "Technical Paper No. 410," *Flow of Fluids Through Valves Fittings and Pipe* (1957).

5

Data Evaluation and Presentation

In chapter 4, I discussed data acquisition. Now that you have acquired all these data, what are you to do with them?

The accumulated database will form the basis for your analysis, conclusions, and recommendations. To get to that point, the data must first be evaluated. Save all that is useful and put aside all that is not pertinent. Don't throw any data away, it may become important. Don't clutter up the analysis with data that are irrelevant either.

When the data have been evaluated and analyzed, the part that is relevant must be organized and presented. This should be done in a way that is simple and clear. A major part of the report writers responsibility is to clarify the data for the client.

EVALUATING DATA

All data must be evaluated for accuracy and applicability to the problem. Accuracy is determined by testing the data against your experience and known fundamentals. For example, a field check may indicate that the design air quantity is being delivered to a space at the design temperature for cooling, but the space is not being adequately cooled. The cooling load must be recalculated (fundamentals) based on the existing usage: people, lights, solar, and equipment. A comparison with the original design will indicate any differences.

Perhaps there will be no significant differences. Then the field data must be rechecked. Instruments may be out of calibration or the test data may be inaccurate. If everything is correct, then there may be other problems, such as controls. At this point, the experience of the investigator becomes very important. Remember, if the problem solution was easy to find your expertise would not have been needed. When the real data have been finalized, an investigation of the air handling system will determine the feasibility of a change in capacity of air flow or temperature. Then, on the basis of your experience, the needed changes can be recommended.

You will note that in the example above both objective and subjective evaluation methods come into play. While an appropriate and correct calculation is objective, the answer you obtain may be a variance with your experience. Experience is mostly subjective. You may know that certain things are correct, but you don't always know why. On the other hand, experience is also objective. Some things simply don't work. You know this because you have tried them under controlled conditions. The professional says, "Yes, I know what my experience tells me, but this new idea may just work and I can't dismiss it out of hand." Don't let experience keep you from accepting new ideas, but remember to examine objectively all new ideas by calculation, experimentation, and trial.

ORGANIZING AND PRESENTING DATA

Before the data can be evaluated and presented, they must first be organized. When I make a simple, obvious statement like that (and you will find several of them in this book) I often wonder why I should insult the intelligence of my readers. Surely everyone is aware of the truth of that comment. Then I remember the many reports I have read that were so disorganized that the writers appeared not to know about simple logic. So I reiterate: organize the data. Arrange it in a logical fashion with related items grouped together. Then you will be able to make a logical analysis and present the results in a logical sequence so that the reader will be informed rather than confused.

Data should be presented in a clear, logical, simple manner. In the summary, only material that is essential to an understanding of the recommendations should be included. In the body of the report, more detail will be needed. Complete calculations should probably be placed in the appendix. Most data can be summarized in tables or figures that allow the major factors to be easily identified. Organizing such a table is not as simple as just listing a number of items. Relationships among the items must also be made apparent. TABLES 5-1 and 5-2 are examples of summary tables.

Table 5-1. An Example of a Summary of Problems.

Problem	Alternative Solutions	Cost	Note
1. Reliability and accuracy	1A. Upgrade existing pneumatic controls	$70,000	
	1B. Replace pneumatic with electronic controls	75,000	Note 6
	1C. Replace pneumatic with DDC (direct digital controls)	133,000	
	1D. Add backup power for control air compressor	2,000	Note 6
2. Documentation errors	2A. Original contractor to correct documents	Note 2	
	2B. Issue contract for corrections	Note 3	Note 6

Problem	Alternative Solutions	Cost	Note
3. Testing of operating sequences	3A. Do not test	No cost	
	3B. Test sequences	Note 4	Note 6
4. Icing of cooling coil, MR.ASU-1	4A. Replace 25-ton compressor with 20-ton compressor	8,000	
	4B. Replace 50-ton condensing unit with 40-ton	20,000	
	4C. Add face damper at cooling coil	2,000	Notes 5&6
	4D. Do nothing	No cost	
5. Condensing Units	5A. Add suction line accumulators, six compressors	5,000	Note 6
6. Building pressure control	6A. Increase minimum outside air		Note 1
	6B. Control outside air with pressure	1,000	Notes 5&6
7. Heat Pump Source water temperature	No change required		
8. Cooling of Cooling Tower Building	8A. Leave unchanged	No cost	Note 6
	8B. Change to two-stage evaporative cooling	5,400	
9. Cooling tower wind eddies	9A. Leave unchanged	No cost	
	9B. Add extension stacks at tower discharge	12,000	Note 6
10. Remote annunciation	10A. Leave unchanged	No cost	
	10B. Extend single point annunciation to control room	3,400	Note 6

Notes:
1. Can be done by staff with engineering support.
2. This should have been done at completion of project and is probably not possible now.
3. Cost about $4,000 if done as part of general revision, about $8,000 if done separately.
4. Cost about $2,000 if done as part of general revision, about $10,000 if done separately.
5. Cost if part of general revision.
6. Recommended alternative.
7. All costs are for construction only, no engineering, design or overhead included.

A presentation can also include figures: graphics, schematics, flow diagrams, and floor plans. Some examples of these are shown in FIGS. 5-1 through 5-4. All tables and figures should be explained in detail in the text, but also should be self-explanatory to the greatest extent

Table 5-2. An Example of a Summary of Alternatives.

	One Plant	Two Plants	Three Plants
Plant Capacity and Load			
Installed capacity, tons	7,200	9,100	11,200
Number of chillers	4	5	7
Annual clg. load, 1000 ton hrs[1]	18,018	18,018	18,018
Operating and Maintenance			
Annual utility costs			
Chillers	$ 427,000	$ 427,000	$ 427,000
Auxiliaries	172,000	172,000	172,000
Total maintenance costs[2]	219,000	226,000	235,000
Total O & M costs	$ 818,000	$ 825,000	$ 834,000
Construction			
Loop integration of existing chillers[3]	$ 446,000	$ 446,000	$ 446,000
Value of existing space[4]	384,000	384,000	384,000
Building service piping[3]	1,248,000	1,248,000	1,248,000
Primary loop piping	8,182,000	6,917,000	6,423,000
CCW Plant[7]	3,230,000	4,310,000	5,520,000
Total Construction Cost	$13,490,000	$13,305,000	$ 14,021,000
Value of existing chillers[5]	2,818,000	2,818,000	2,818,000
Construction Cost of New Plant (20-year life)	$16,308,000	$16,123,000	$16,839,000

Notes:

1. Values developed in Tables VII-6 through VII-8.
2. Values from Table VII-5.
3. Values from Tables VII-1 through VII-3.
4. Assuming 2,000 square feet per chiller unit at $60 per square foot for space occupied in buildings.
5. Remaining value of existing chillers computed on 20 year straight-line depreciation and unit cost per ton of capacity. See Table VII-9.
6. Btu meters at buildings using 100 tons or more will add approximately $46,000 to any scheme (connection of meters to EMCS is not included).
7. Values from Table VII-4.

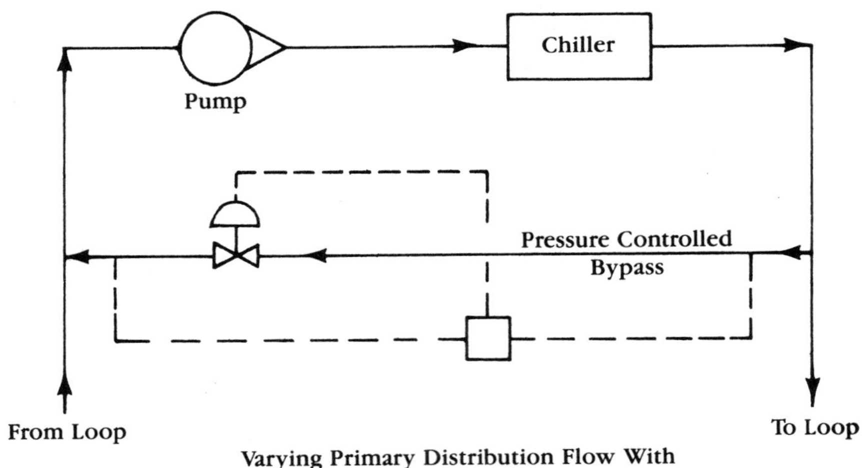

Varying Primary Distribution Flow With
Constant Chiller Flow—Pressure Controlled Bypass

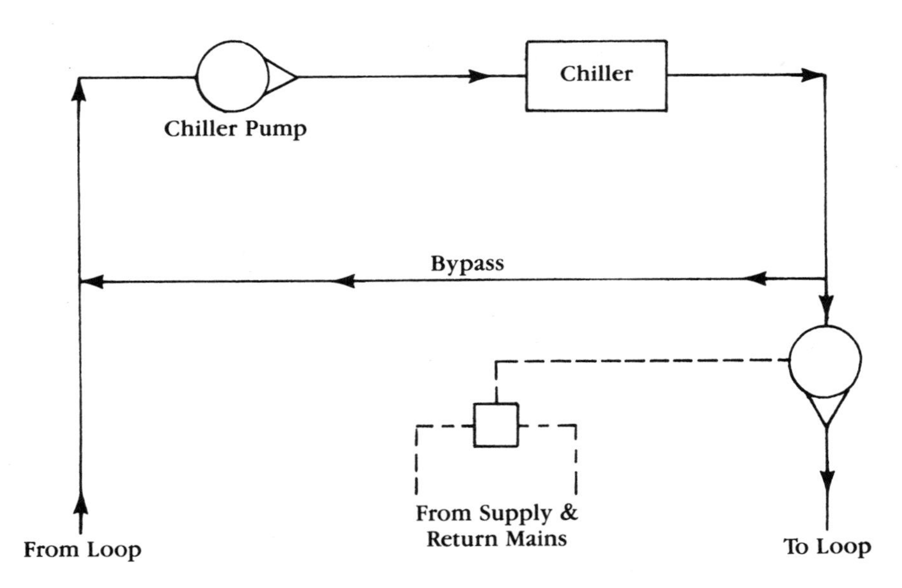

Note: ΔP to maintain constant pressure differential between supply and return mains at selected point.

Varying Primary Distribution Flow With
Constant Chiller Flow—Variable Speed Pump

Fig. 5-1. Examples of schematic flow diagrams.

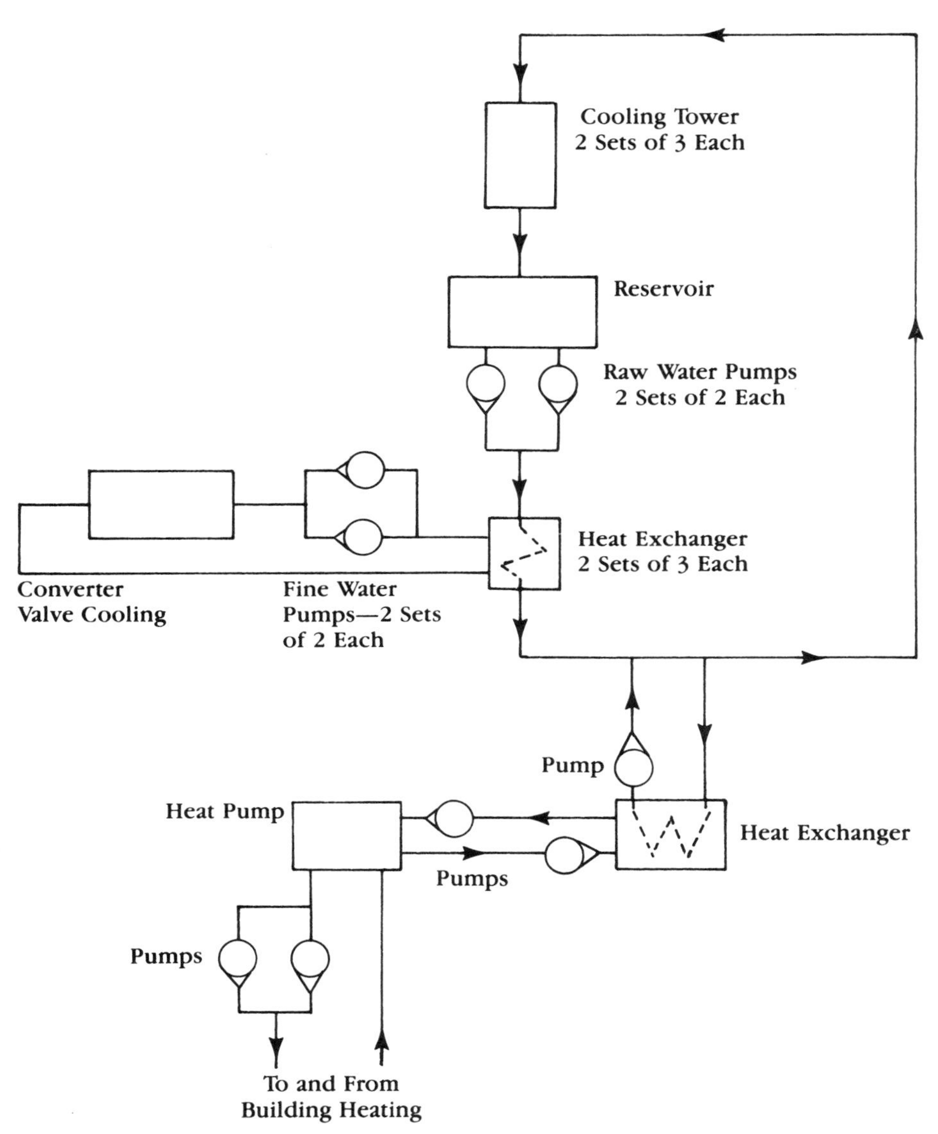

Fig. 5-2. An example of a schematic flow diagram.

possible. TABLES 5-1 and 5-2 show how notes can be used for clarification.

When preparing a presentation, it is necessary to remember that the reader does not have the advantage of

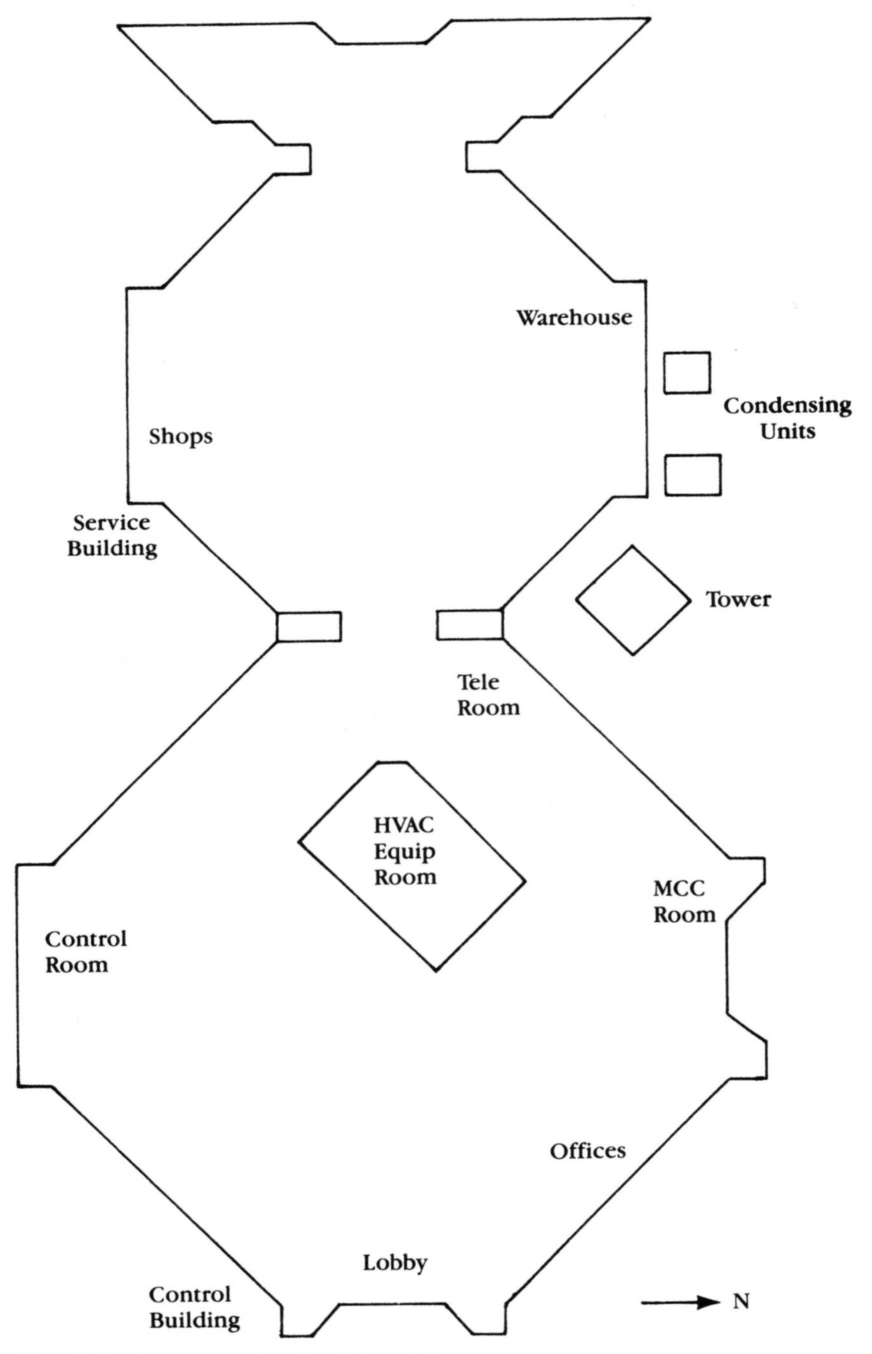

Fig. 5-3. An example of a floor plan.

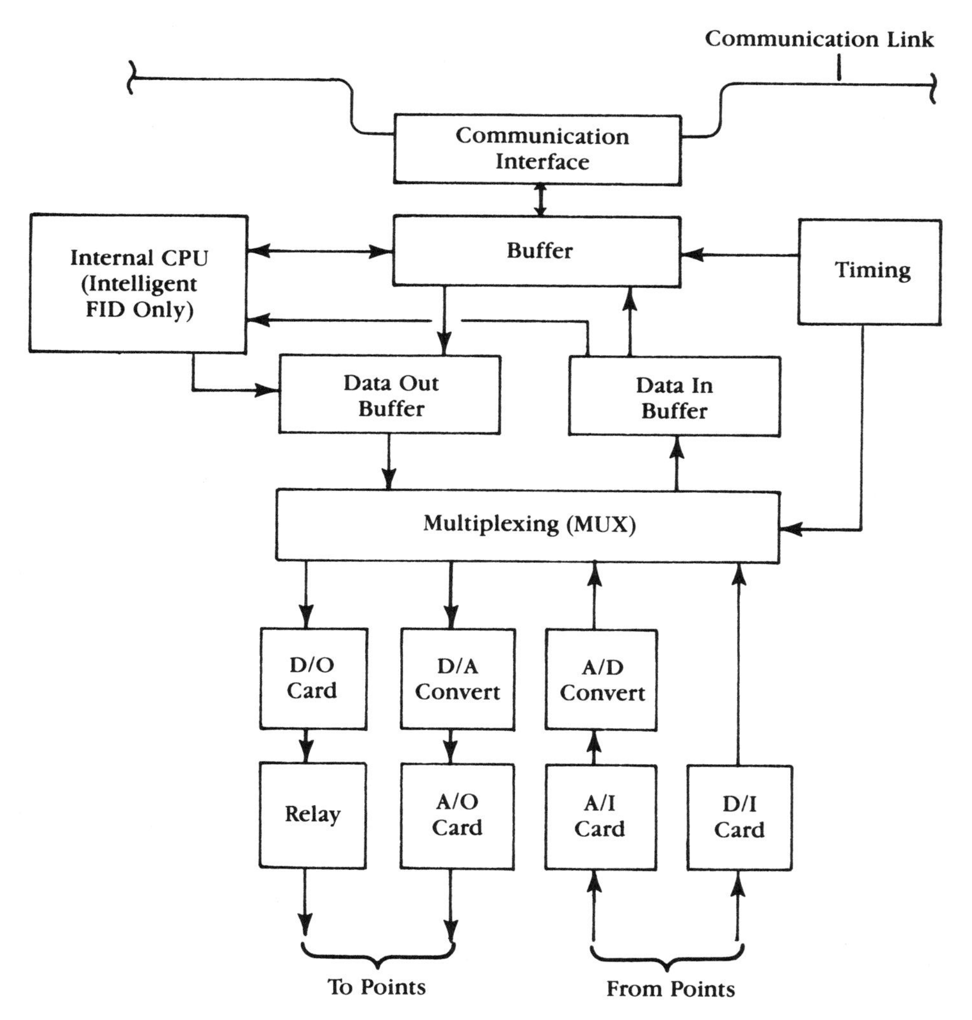

Fig. 5-4. An example of a schematic arrangement.

your experience nor the in-depth knowledge you gained from studying the problem. You have to start with fundamentals and provide enough background so that a reasonably educated reader can follow the discussion. It is best to assume that nothing is obvious. Never say "of course," "it is obvious," "it is apparent," or "it is well known that." None of these may be true for the reader.

EXPLAINING DATA AND RECOMMENDATIONS

No matter how detailed, logical, and complete the presentation may be, it is incomplete without an explanation of what it all means to the client. *Explanation* refers to such statements as: Alternative No. 1 is more complex than the others, but will provide the required environmental control; or, These data indicate that the system is not operating as designed. Follow these statements with specific examples. Explanations should help the client make an informed decision, which, in the end, is the purpose of the report.

Explanations of meaning should relate primarily to the problem needing a solution. For example, if the problem is one of reliability, then the explanations should be concerned with reliability. They should explain how each alternative addresses that issue. The client also needs to know how well each alternative will solve the problem. A cheaper alternative may not provide the best solution, but it may be sufficient for the client.

Explanations should be as objective as possible—reports are supposed to be objective. You must realize, however, that clients will not necessarily be objective in their evaluation of the report. The client frequently does not tell the report writer about his or her hidden agenda, which may include a great many subjective factors such as personal likes and dislikes, budget problems, etc. The report writer needs to recognize that not all report recommendations will be implemented and that this lack of implementation does not necessarily mean that the report writer has failed.

6

Alternatives and Recommendations

It is not sufficient to present recommendations without explanations. Recommendations should be presented as the preferred selection of several alternatives. Each of the alternatives also should be given consideration, with its merits and demerits defined. From this information, it should be clear to the client that the recommendation is based on a rational, logical analysis, rather than simply the preferences of the writer. It also allows the client to select another alternative other than the one recommended when he considers it in his best interest to do so.

DEVELOPMENT OF ALTERNATIVES

Alternatives are developed out of the problem statements and the available data. Any alternative must provide a solution of some sort. Not all proposed solutions will be viable, but all possible solutions should be examined. The process called brainstorming is helpful here. It is often helpful to look at the problem in nontechnical ways. The knowledge of engineering limitations may sometimes limit your thinking. At this point, no evaluation should take place. You should be interested only in listing all possible solutions.

The experience of the investigator is often the best source of ideas. It is useful to know, from experience, what has been effective before in similar situations. Experience also can be detrimental because the investigator knows what didn't work before and may not recognize that this time there are subtle but important differences. The opposite side of this argument is that what worked elsewhere may not work in this situation, at least not without some modifications.

Another excellent source of ideas is the comments of the facility operators and, to a lesser extent, the building occupants. Again, you need to remember that the information obtained from the operators is, to a very great extent, dependent on your approach. If you treat the operator with respect and listen carefully to his suggestions, he will be much more willing and able to provide good information.

Often, the client will have an agenda and suggestions. Since he is paying for your services, it is required that you pay attention. His suggestions may be satisfactory, if not the best, solutions. They should be included and evaluated. As previously noted, the client may select those alternatives anyway, regardless of your recommendations.

ANALYSIS AND EVALUATION OF ALTERNATIVES

When all the alternatives have been listed, it is time to proceed with analysis and evaluation. There are several criteria that can be applied to this process (TABLE 6-1).

Are the technical requirements satisfied? Any solution that does not meet the technical requirements is not a solution. The real question here is how well those requirements are met. A manufacturing process with a tight environmental specification leaves little room for discussion. A comfort criterion is broad enough for considerable interpretation. In both situations, the energy conservation aspect of each alternative should be considered.

What is the level of complexity? Complex systems probably cost more to install. They are certainly more difficult and expensive to operate. If the complexity seems necessary, then the proposed solution must include the training that will allow the operators to use the system in the way it was designed. If this is not done, the system will be operated in a simpler manner that probably will not be satisfactory. The simplest is the best rule may not always hold, but it should be considered.

Is the proposed alternative practical? Can the equipment actually be installed that way? Is there space available not only to install it, but to operate and maintain it? Can the modification be made without a major or unacceptable shutdown of the clients operations? Some manufacturing plants allow only one or two shutdowns per year. All major modifications must be planned around that criterion. Some facilities, such as telephone exchange buildings, cannot be shut down at all.

Are standard components used? Special designs may be needed, but are always more expensive. Look for ways to use off-the-shelf equipment. What level of quality does the alternative require? Lower quality equals lower first cost, but may increase operating costs and prevent the achievement of the desired results.

Is the system acceptable to the operating and maintenance people? They are the ones who will have to live with it. Can they make it work, given the opportunity? Or, can they make it fail if they don't like it? Any proposal must consult, educate, and motivate these people. It is often necessary to remind management of that fact.

Table 6-1. Criteria for Evaluating Alternatives.

1. Are the technical requirements satisfied?
2. What is the level of complexity?
3. Is the proposed alternative practical?
4. Are components standard or special?
5. Is the system acceptable to the operating and maintenance people?
6. Is the cost reasonable?

Is the cost reasonable? In other words, is the cost within the original budget? If not, there must be a good reason for the increase. Perhaps the original budget was unrealistic in view of the technical requirements. Cost covers a great many things. First cost is only the beginning. The cost of education and maintenance increases exponentially with the increased complexity of the system. Life cycle cost analyses must often be made in order to properly compare alternatives.

SELECTION AND PRESENTATION OF ALTERNATIVES

When the analysis and evaluation have been completed, the better alternatives will become apparent. There is seldom any one best solution to a problem. Instead, the solutions tend to be good and better. All viable solutions should be presented. The reasons for selecting solutions can be of assistance when you make the final recommendations. The desires of the client also play a part, since any alternatives to which he is partial must be considered and included if possible. If the client's solutions are not part of the better group, the analysis should clearly show the reasons.

The selected alternatives should be listed along with their benefits and drawbacks. In addition to the text discussion, it is very useful to present these conclusions in tabular form (TABLE 6-2).

RECOMMENDATIONS

The recommendations should arise out of the selection of alternatives. Each recommendation must always have its basis in the analysis. Sometimes the recommendation is obvious. At other times, two or three alternatives may appear equal. The report writer's experience should lead him to a preferred alternative, but the report should also give the other alternatives that will provide some level of satisfaction. The recommendations must be stated in the

Table 6-2. An Example of a Table of Alternatives.

Description	Design and Acquisition Cost[1]	Annual Savings Energy, Mbtu[2]	Annual Savings Cost[3]
1. Replace existing windows with double-pane insulating windows with solar gray exterior glass.	$130,000	184,000	$ 950
2. Replace existing windows with thermal-break insulating windows. Replace existing curtain wall.	305,000	237,500	1,220
3. Install storm windows with operable sash.	250,000	190,500	980
4. Insulate and reslope roof for better drainage.	180,000	31,900	165
5. Replace old roof with new roofing system.	190,000	31,900	165
6. Insulate exterior masonry and curtain wall. Provide new interior finish.	305,000	321,500	1,650
7. Reduce the number of usable windows by 25% by insulating and furring.	9,200	45,100	232

Notes:
1. Costs include demolition, materials, labor, and design.
2. Mbtu = thousands of Btu's per year.
3. Costs based on boiler efficiency of 70% and current gas costs of $3.60 per 1,000 cubic feet.

text, but can also be included in a tabular presentation of alternatives (TABLE 6-3).

SUMMARY

While the alternatives and recommendations section of the report is the most important part from the client's viewpoint, it should result naturally from work you have done in order to acquire, organize, and analyze the data. This section should be the easiest part of the report to write if you use the same clear, logical approach you have used throughout.

Table 6-3. Tabular Summary of Alternatives and Recommendations.

Problem	Solution	Budget Cost[1]
1. Balancing outside air makeup and exhaust air	a. Increase outside air capacity of units ACS-1 and 4	$54,000
	b. Preheat/precool outside air, old building only.	90,000
	c. Provide a common mixed air plenum, both buildings	41,000
2. Total supply air versus exhaust plus return	a. Convert return air fans to VAV	8,000
	b. Provide balancing dampers in return air ducts	16,000
	c. Rebalance return air, increase capacity of old building return air fan	9,000
	d. Increase air flow rates of HVAC units ACS-3, 4 & 5	17,000
	e. Provide transfer ducts where needed	11,000
	f. Increase capacity of new building return air system	42,000
3. Controls	a. Provide programmable controller, including wiring and software	70,000

Notes

1. Budget costs are detailed in Section IV. They include normal construction costs but no design costs or related costs, such as interference with building operations.
2. Recommended solutions include 1c plus all of 2 and 3 for a total of $214,000.

7

Graphics

In a report, graphics are the figures, tables, summaries, and photographs that illustrate and dramatize the text. Someone once said, "One picture is worth a thousand words." This is certainly true for graphics. Since clarity and simplicity of presentation are necessary in order to get a message through to the busy manager, a properly organized summary table, graph, or figure can be the key to a successful report. While even a poor graphic may be useful, good quality is important to credibility.

TABLES

Almost any kind of data can be presented in tabular form. Tables may summarize equipment, alternatives, recommendations, or costs. TABLE 7-1 is a summary of alternatives and recommendations from heating, ventilating, and air conditioning controls study report. TABLE 7-2 is a cost summary from a ventilation system study report. Each of these tables was used in section I, the report summary. Each is based on tables and data from other sections of the report. Both use footnotes for clarification. The summary table should stand on its own, even though the text is necessary for a full understanding of all the assumptions and reasoning behind the data.

Table 7-1. Summary of Problems, Alternative Solutions, and Recommendations.

Problem		Alternative Solutions	Cost	Note
1. Reliability and accuracy	1A.	Upgrade existing pneumatic controls	$70,000	
	1B.	Replace pneumatic with electronic controls	75,000	Note 6
	1C.	Replace pneumatic with DDC (direct digital controls)	133,000	
	1D.	Add backup power for control air compressor	2,000	Note 6
2. Documentation errors	2A.	Original contractor to correct documents	Note 2	
	2B.	Issue contract for corrections	Note 3	Note 6
3. Testing of operating sequences	3A.	Do not test	No cost	
	3B.	Test sequences	Note 4	Note 6
4. Icing of cooling coil, MR.ASU-1	4A.	Replace 25-ton compressor with 20-ton compressor	8,000	
	4B.	Replace 50-ton condensing unit with 40-ton	20,000	
	4C.	Add face damper at cooling coil	2,000	Notes 5&6
	4D.	Do nothing	No cost	
5. Condensing Units	5A.	Add suction line accumulators, six compressors	5,000	Note 6
6. Building pressure control	6A.	Increase minimum outside air	Note 1	
	6B.	Control outside air pressure	1,000	Notes 5&6
7. Heat Pump Source water temperature		No change required		
8. Cooling of Cooling Tower building	8A.	Leave unchanged	No cost	Note 6
	8B.	Change to two-stage evaporative cooling	5,400	
9. Cooling tower wind eddies	9A.	Leave unchanged	No cost	
	9B.	Add extension stacks at tower discharge	12,000	Note 6
10. Remote annunciation	10A.	Leave unchanged	No cost	
	10B.	Extend single point annunciation to control room	3,400	Note 6

Notes:

1. Can be done by staff with engineering support.
2. This should have been done at completion of project and is probably not possible now.
3. Cost about $4,000 if done as part of general revision, about #8,000 if done separately.
4. Cost about $2,000 if done as part of general revision, about $10,000 if done separately.
5. Cost if part of general revision.
6. Recommended alternative.
7. All costs are for construction only, no engineering, design or overhead included.

FIGURES

Figures include schematic drawings graphs, pie charts, and bar charts, maps, floor plans and elevations, and

Table 7-2. An Example of a Cost Summary.

Scheme	Description[1]	Installation Cost[2]
1	One new central A/C & makeup for building with heat reclaim New central exhaust	$494,000
2	Separate A/C & makeup for lab & office with heat reclaim New central exhaust	535,000
3	Leave existing A/C New central makeup air New central exhaust with heat reclaim	438,000
4	New central A/C with heat reclaim New central makeup air New central exhaust	520,000
5	New central A/C Leave existing A/C and exhaust	339,000

Notes

1. All systems include VAV controls and all except Scheme 5 include heat reclaim.
2. Cost includes air systems, new chiller and boiler, new equipment room, demolition of existing equipment and ducts as applicable, removal and replacement of office ceilings, and new ceiling in labs.
3. Costs include installation of new ceilings in the laboratories. The added cost for the new lay-in ceilings in the laboratory areas would be approximately $7,000. If the new supply ducts were hung beneath the existing ceiling, the installation costs would be reduced by approximately $20,000.

illustrations. Figure 7-1 shows how a psychometric chart can be used to illustrate a cooling/dehumidifying cycle. Figure 7-2 is a schematic drawing that shows two methods of controlling flow rates in a chilled water system.

Schematic drawings of systems and controls are common. Isometric and perspective illustrations are used to show the client how the system will appear when installed. Floor plans and elevations are needed to show relationships within a building. Maps show relationships between buildings. Graphs of various kinds are used to show other relationships and to make comparisons among the various elements, especially costs, in the study.

Figures should be neatly and carefully drawn by a competent draftsperson or a professional illustrator. They also may be produced on a computer aided drafting (CAD) system or by a simple computer graphics program.

For hand-drawn work, either freehand or machine lettering is acceptable. Several types of machines are available. One widely used machine produces a printed stick-on tape. Several type styles and sizes are available. Another system uses rub-off letters, again with many available sizes and styles. Stencils and lettering guides can be used as in FIG. 7-2. The lettering in FIG. 7-1 was typed on plain, stick-on mailing labels, which were then trimmed to size and placed on the drawing.

Many computer systems include lettering. It is important that the lettering be neat and legible. It must also clarify elements of the drawing. Freehand lettering is generally the quickest, but it requires a degree of talent that some of us may lack. It is preferable to use machine or computer methods. Large figures may need to be reduced to fit on the report page. Lettering should be large enough that it can read easily after reduction.

Rough, freehand sketches may sometimes be used along with calculations in the appendix, but never in the body of the report.

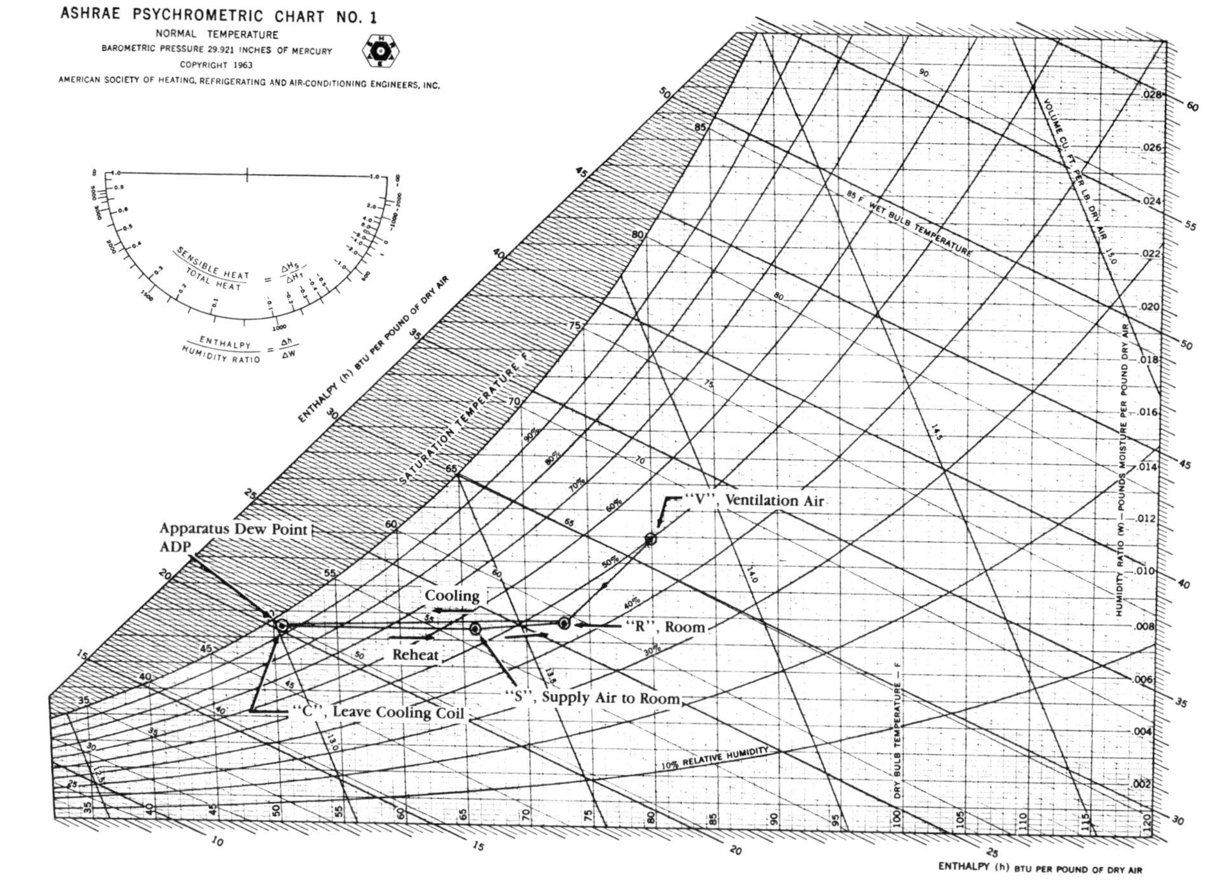

Fig. 7-1. An example of a psychometric chart.

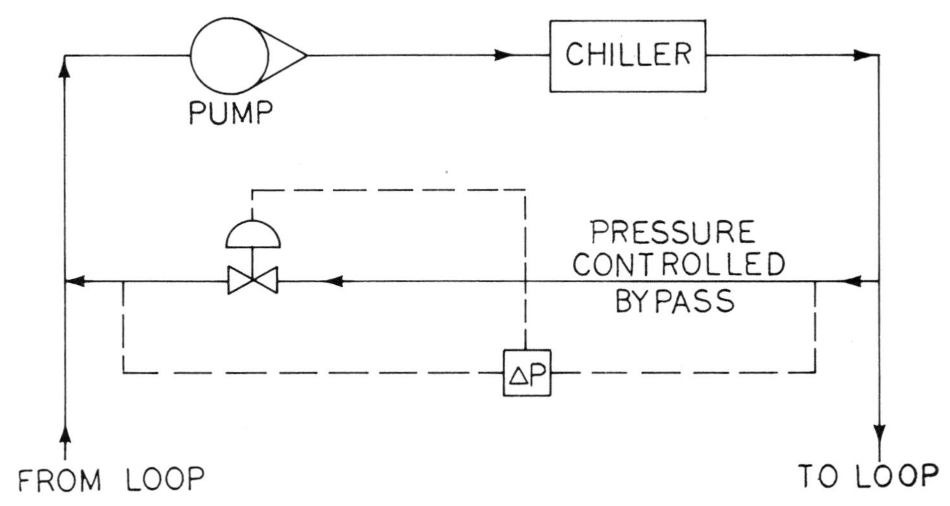

VARYING PRIMARY DISTRIBUTION FLOW WITH
CONSTANT CHILLER FLOW - PRESSURE CONTROLLED BYPASS

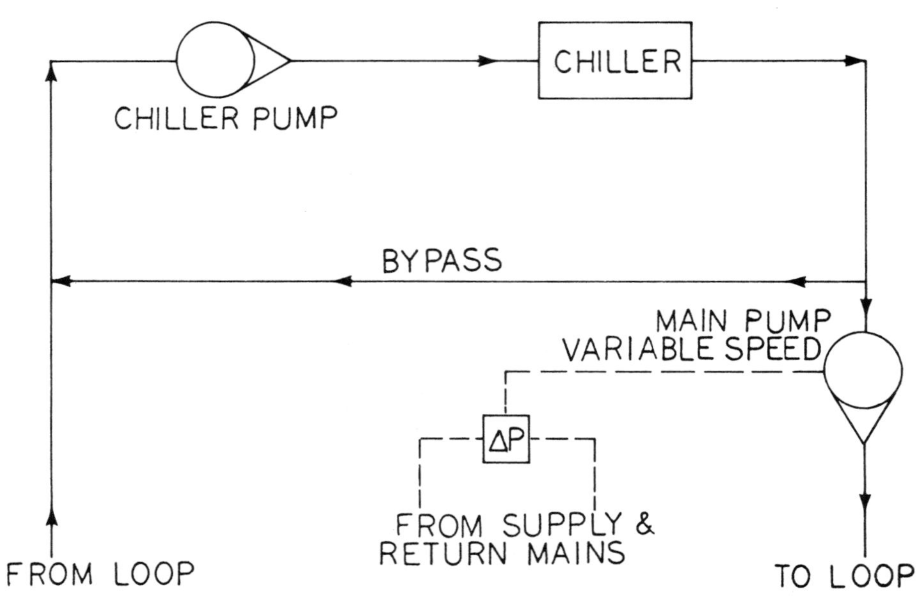

Note: ΔP to maintain constant pressure differential between supply and return mains at selected point.

VARYING PRIMARY DISTRIBUTION FLOW WITH
CONSTANT CHILLER FLOW - VARIABLE SPEED PUMP

Fig. 7-2. An example of a chilled water schematic chart.

PHOTOGRAPHS

Photographs are useful for showing existing conditions. They are more expensive to reproduce, however, since ordinary copying or printing methods are not satisfactory. Clarifications or proposed modifications can be marked on a photograph. This technique has also been used on design documents. Often it is a better way of improving the understanding of the situation than an ordinary drawing.

COLOR AND SHADING

Color is not normally used in technical reports, primarily because of the cost and difficulty of reproduction. Since color is used so widely in sales and promotional literature, there may even be some resistance to its use in technical reports on the basis that a report is supposed to avoid emotive material. However, it can be useful in clarifying graphics.

There are other methods available for providing contrast and emphasis. The most common is the use of stick-on, patterned shading and cross-hatch overlays. These are available in many variations and patterns. They are applied to precise outlines by cutting and trimming with a razor knife as in FIG. 7-3. This is faster and creates a better appearance than manual shading. The stick-on material provides the half-tone shades necessary for most reproduction methods. This type of shading can be added to computer drawings, though the computer system may also be able to generate its own crosshatching.

COMPUTER GRAPHICS

If a CAD system is available, it may be used. Smaller, less expensive, and simpler computer systems and programs also can be used to produce report-size graphics. Graphics programs are available for most personal computers. The quality of reproduction varies somewhat with the system and especially with the type of printer used.

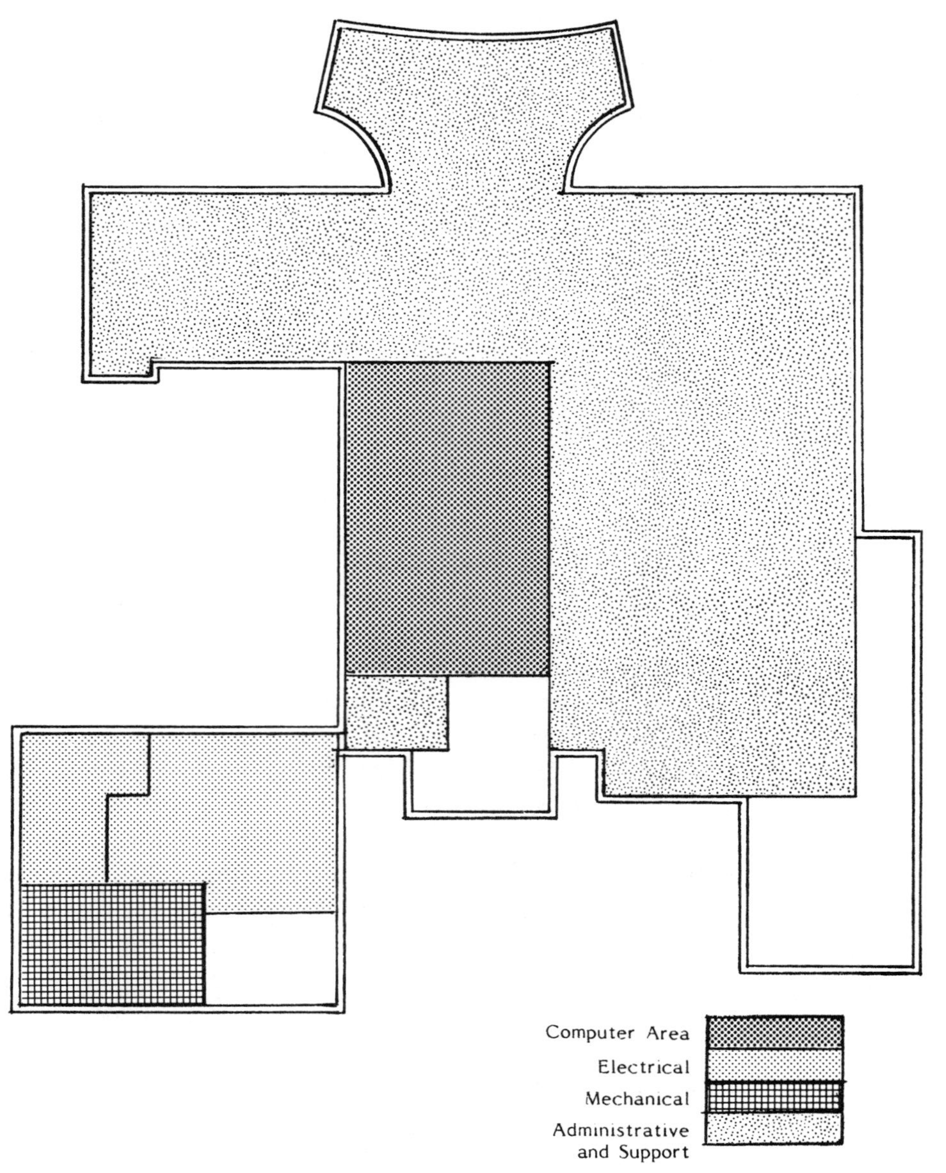

Fig. 7-3. An example of half-tone shading.

The oldest and most common system is the dot-matrix printer (Fig. 7-4). The name describes the printing method: all characters and lines are printed using a matrix of small ink dots. Modern dot-matrix printers provide high quality copy, though the dot separation is still visible. These printers can produce almost any graphics desired. They can also be used to produce text when combined with word processing programs.

Plotters, which are actually small CAD systems, use an X-Y plotter, to produce the graphic. The system also includes lettering. One difficulty with this system is the drafting of circles and arcs, which must be produced by a series of short vertical and horizontal lines. Plotters are more expensive than dot-matrix printers.

Daisy wheel printers (FIG. 7-5) provide high-quality, typewritten text with features such as right justification.

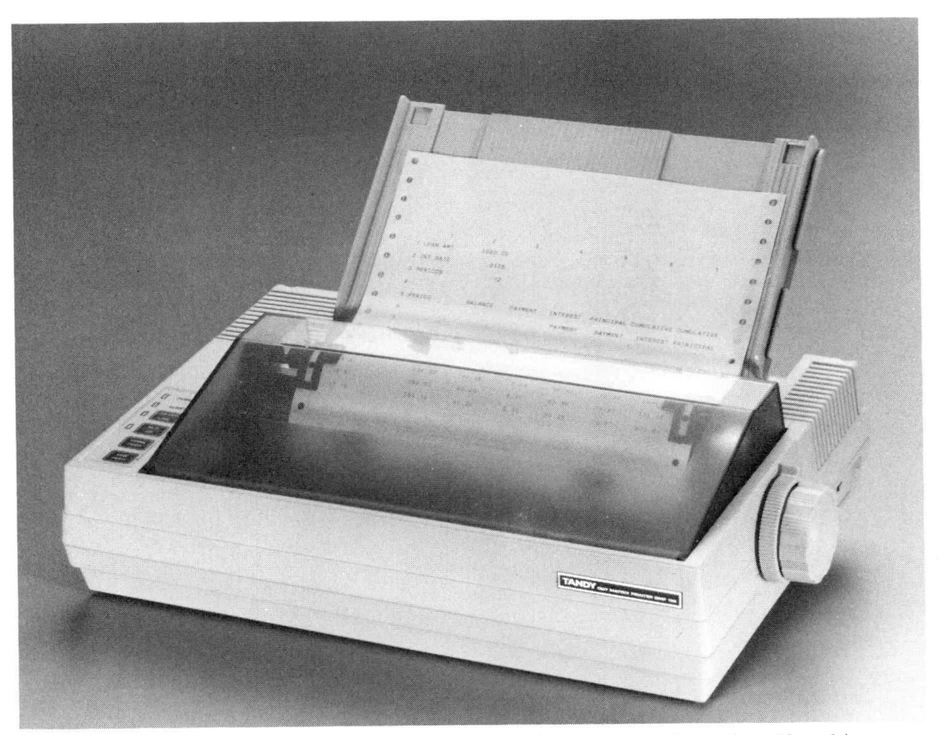

Fig. 7-4. A photo of a dot-matrix printer (courtesy of Radio Shack).

Many different type fonts are available, but you must physically change the print wheel in order to change the type face. Daisy wheel printers do not do graphics.

The newest and best systems use a laser printer (FIG. 7-6) for reproduction. This system, with the proper software, will provide a camera-ready copy of any text or

Fig. 7-5. A photo of a daisy-wheel printer (courtesy of Apple Computer, Inc.).

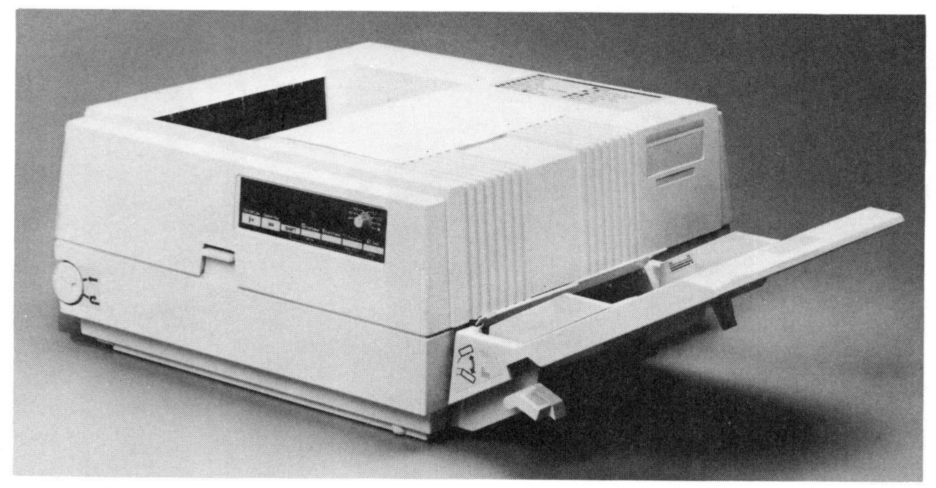

Fig. 7-6. A photo of a laser printer (courtesy of Apple Computer, Inc.).

§ *101 LEGISLATION/STATUTES*

habeas corpus petitioner. *Parks v. Christensen*, 751 F.2d 1081 (9th Cir. 1985).

Statutory construction. (101) When interpreting a statute, the court will not go beyond its language unless it is ambiguous or rendered so by other inconsistent statutory language. *U.S. v. Roach*, 745 F.2d 1252 (9th Cir. 1984).

Standard of review, statutory construction. (101) The *de novo* standard of review applies to questions involving construction of statute. *U.S. v. Wilson*, 720 F.2d 608 (9th Cir. 1983).

Existence of more specific statute did not bar conviction under general false claim statute. (101) It was proper to convict the defendant under the general false claim statute, 18 U.S.C. Section 287, even though his false claims for Supplemental Security Income were specifically proscribed by the misdemeanor false claim provisions of 42 U.S.C. Section 408. *U.S. v. Ruster*, 712 F.2d 409 (9th Cir. 1983).

Constitutional Issues, Generally §130

Supreme Court upholds constitutionality of new federal sentencing guidelines. (130) In an 8-1 decision written by Justice Blackmun, the Supreme Court upheld the new federal sentencing guidelines against a claim that they violated the doctrine of separation of powers. The decision effectively overrules the Ninth Circuit's contrary opinion in *Gubiensio-Ortiz v. Kanahele*, 857 F.2d 1245 (9th Cir. 1988). The Supreme Court upheld Congress' placement of the Sentencing Commission in the judicial branch, and found no flaw in the requirement that federal judges serve on the Commission and share their authority with non-judges, nor in the fact that the President appoints members of the Commission and may remove them for cause. The court also rejected the argument that Congress delegated excessive legislative power to establish sentences to the Commission. Justice Scalia dissented at length. *Mistretta v. U.S.*, __ U.S. __, 109 S.Ct. 647 (1989).

Statute authorizing death penalty where murder was "especially heinous, atrocious or cruel" was unconstitutionally vague. (130) In a unanimous opinion written by Justice White, the Supreme Court held that an Oklahoma statute authorizing the death penalty to be imposed where the murder was "especially heinous, atrocious or cruel" was unconstitutionally vague. An ordinary person could honestly believe that every unjustified, intentional taking of human life is "especially heinous." Thus these words failed to guide the jury's discretion in imposing the death penalty. The court returned the case to the Oklahoma courts for resentencing, noting that after this case was decided, the Oklahoma Court of Criminal Appeals had construed the phrase to require torture or physical abuse, and had indicated that where, as here, a separate valid aggravating circumstance was found, it would not necessarily set aside the death sentence. *Maynard v. Cartwright,* __ U.S.__, 108 S.Ct. 1853 (1988).

Statute prohibiting picketing within 500 feet of a foreign embassy violates First Amendment. (130) Congress enacted, as part of the District of Columbia Code, a statute prohibiting the "display" of any sign within 500 feet of a foreign embassy that tends to bring that foreign government into "public odium" or "public disrepute." Justice O'Connor, writing for a 5-3 majority, held that the statute violated the First Amendment. The majority noted that 18 U.S.C. section 112, which applies everywhere except the District of Columbia, prohibits acts intended to "intimidate, coerce, threaten or harass" foreign embassies, and that the availability of such alternative statutes demonstrates that this statute is not "crafted with sufficient precision" to withstand First Amendment scrutiny. However another portion of the statute, which prohibited any "congregation" of three or more persons within 500 feet who refuse to disperse after a police order to do so, was upheld on the ground that it had been sufficiently narrowed by interpretation by the

NINTH CIRCUIT CRIMINAL LAW REPORTER 4

Fig. 7-7. An example of laser printer output (courtesy of La Jolla Legal Publications).

graphic material. Laser printers are relatively expensive, but their versatility and high quality make them worth while for those who do a large volume of report work. The laser printer also is used to produce text. It has a virtually unlimited range of type styles and size (FIG. 7-7).

8

Writing a Summary

The summary, as its name implies, summarizes the report's findings in a succinct, complete manner. It is written for the purpose of allowing the reader to gain a general understanding of the information in the report in a short time. For those readers who are not interested in the details, the summary provides the minimum background necessary to understand the conclusions and recommendations in the report. For those who are interested in and will read the detail, it provides an overview that allows the reader to keep the information in the report in perspective. For these reasons, the summary must be complete, but include only essential information.

The summary should be the last part of the report you write. If you find yourself writing the summary, even mentally, before the detailed work has been done, you are in trouble. It takes mental discipline to avoid early summarizing, but it is essential to a well-written report.

HOW TO SUMMARIZE

When I was in college I took a required course called Writing Scientific Papers. One of the exercises required of the students was to reduce any piece of writing by a factor of 10 without losing the essential meaning. That meant 10 words of summary per 100 words of original

Table 8-1. Summary of Objectives.

I. REDUCE OPERATING COSTS
   A. Reduce Energy Consumption and Demand
   B. Reduce Labor Costs
II. IMPROVE SYSTEM OPERATION
   A. Simplify and Standardize HVAC Systems and Controls
   B. Provide Early Response to Operating Problems
   C. Improve reliability of Systems
   D. Provide Maintenance Scheduling
   E. Provide Good Historical Operating Data

writing. It was an interesting and instructive exercise. It proved to me that the essential idea could be conveyed very simply. Most of what you write is simply exposition—an attempt to clarify an idea. Sometimes the extra material is helpful, especially illustrations and examples. Sometimes the extra material hinders communication.

In report writing, the development of detailed background information and data, analysis and evaluation, and conclusions and recommendations takes a great many words. There is nothing wrong with this. It is necessary for the writer and the reader to see all the steps you go through when you prepare the study. The summary must reduce this mass of detail to a few basic facts. The ratio of 10 words of summary to 100 words of detail is not at all unreasonable.

Detailed writing, logically done, contains a key sentence in each paragraph. The rest of the paragraph develops the key idea. The summary should include the key sentences, either copied exactly or paraphrased. If the detailed sections of your report lack this logical structure, then you need to do some rewriting!

There also will be tables or graphics in the detailed sections of your report that summarize information. These can be reused or further summarized in the report summary (TABLE 8-1).

Various alternatives will also need to be summarized in statement form without the supporting data provided in the body of the report. The recommendations should be summarized in the same way. For both of these items, the use of summary tables is very helpful (TABLE 8-2).

Table 8-2. An Example of a Summary Table of Problems and Solutions.

Problems	Solutions
1. Poorly designed or maintained buildings.	1a. Redesign: add insulation, weatherstripping, etc. 1b. Improve maintenance.
2. Inefficient occupancy patterns, partial use of buildings.	2a. Make administrative changes to minimize partial occupancy. 2b. Redesign HVAC systems for greater flexibility of use.
3. Inefficient HVAC systems and/or control systems.	3a. Redesign, replace, or retrofit HVAC systems for greater efficiency. 3b. Retrofit control systems to improve accuracies and strategies.

A letter report is essentially a summary. Because there is no following, supporting information, however, it must include more details. In the letter report, the basic reasoning behind the alternatives and recommendations must be included. When you prepared the letter report, you developed more detailed data than you put into the short report. It is this detail that should be summarized.

WRITING FOR A SPECIFIC AUDIENCE

As already noted, the audience for the report consists of two groups. The first is the management group, which is interested primarily in the financial impact of the recommendations. The members of this group also need to be concerned with the costs of staffing and training, but often are not aware of this need. The summary must address their needs and interests.

The second audience is the technical group: those who use, operate, and maintain the systems. Their interest is in the usability and technical simplicity of the recommendations. They are not always aware of the inherent staffing and training problems. The technical group is interested in the financial aspects to the extent that budgets are not exceeded. They are usually very aware of the energy and operating costs implied by the recommendations.

Throughout the writing of the report, but especially in the summary, you must be aware of the needs of these two groups. Both groups must be served. This is not always easy, but it must be done if the report is to be successful.

SUMMARY

The summary must be complete, yet succinct. That is not as simple as it sounds. The report writer must learn to reread his summary with a great deal of objectivity. I usually try it on a colleague or my wife. Remember that the reader probably doesn't know as much about the subject as the writer. If he does, the report isn't needed.

9

The Appendix to the Report

The appendix to the report should contain reference material that provides details and information, which support the body of the report. Appendix material may include a glossary of terms, symbols, and abbreviations; detailed calculations; dissertations on topics alluded to in the report; references and citations; and anything else that the report writer feels is desirable, but prefers not to put in the main sections of the report.

WHEN TO USE AN APPENDIX

The decision to use an appendix is subjective. In general, any lengthy report will need an appendix. A short, simple report will not. Between these two extremes some criteria are needed. Any material that is needed to verify the report conclusions must be included somewhere. If it makes the analytical discussion cumbersome, then it should be in the appendix. If it amplifies the discussion but is not essential to the argument, then it should be in the appendix. Detailed calculations and cost analyses are sometimes placed in separate sections or they may be included in the appendix.

Table 9-1. A Sample Glossary.

Algorithm: A description of the steps or process in the solution of a problem.
Analog: Data in continuous form. An analog signal can take on any value within the range of a control instrument.
AI: Analog In—an analog signal from a sensor.
AO: Analog Out—an analog command signal to a controlled device.
bit: A single data item in a digital computer that has a value of one or zero.
byte: A group of eight bits.
controlled device: In a control system, the final device that receives the output of the controller, typically a damper or valve operator, or motor.
communication system: The wires or cables over which a computer may talk to another computer or to peripheral and interface devices.
CPU: Central Processing Unit—the central computer in an MCS, sometimes called Host Processing Unit (HPU).
CRT: Cathode Ray Tube—an electronic vacuum tube containing a screen on which information may be displayed in graphic or alpha-numeric form.
data: Information.
database: The list of system points, with information about them, that is stored in the memory of a CPU, IFID, or DDC computer.

CONTENT OF THE APPENDIX

Some of the more common items included in the appendix are discussed below.

Glossary. Most reports will need a glossary of terms, abbreviations, and symbols. An example is shown in TABLE 9-1. If the list is short, it may be included at an appropriate place in the body of the report. For most reports, it will be more reasonable for it to appear in the appendix. The glossary will include definitions of technical terms, abbreviations, and acronyms, even though they may also be defined when first used in the report. It is particularly difficult for the uninitiated to remember acronyms. For figures such as control diagrams, a list of the symbols used must appear somewhere, preferably in the appendix.

Calculations. Calculations that demonstrate the accuracy of the analysis must be included. The inclusion of detailed calculations in the body of the report may be counterproductive. Most readers tend to skip over detailed calculations. In any case, the thread of the discussion may easily be lost while trying to verify the integral of some function. It is more effective to state "It can be shown that," and then refer those who wish to check the mathematics to the appendix. (FIG. 9-1.)

References and citations. In the preparation of your report, you will refer to many sources of information. It is helpful to the reader and increases the credibility of your conclusions if these references are cited. The typical format for these references has been used at the end of each chapter of this book to cite research material I used. References may be included at the end of each section or in the appendix.

Dissertations. By definition, technical reports discuss technical subjects. As a report writer, it often appears that your audience may need further education in the topic under discussion. In the appendix, you can include a dissertation on the topic. For example, most engineers are not fully cognizant of the differences in modulating control modes—proportional, integral, and derivative—so I frequently include a short discussion of the meaning and importance of these modes in reports on control systems.

THE VALUE OF THE APPENDIX

The usefulness and value of the appendix can be great if it is used properly. The summary is simple and short for those who have limited time to give to the project. The analysis goes into detail for those who are interested in the technical aspects and wish to verify the conclusions. The appendix not only provides additional verification, but offers the opportunity for further study to those who wish to become better educated.

The appendix is of value to us as writers because it describes another level in the depth of our analysis, thereby improving our credibility. Even when we are

DETERMINING STACK EFFECT ΔP IN XYZ BUILDING

11-15-89
RWH -
JOB. 902

REF: ASHRAE FUNDAMENTALS, 1989.
PAGE 23.4, EQUATION 9

$$\Delta P_s = C_z d_i g (h - h_{NPL}) \left(\frac{T_i - T_o}{T_o} \right)$$

C_z = UNIT CONVERSION FACTOR = 0.00598

d_i = air density ≅ 0.075 lb/ft³

g = gravitational constant = 32.2 ft/sec²

h = height of observation ft., assumed 5 ft, (median of 2nd basement)

h_{NPL} = height of neutral pressure level, ft.

T = absolute temp, °R, Assume $T_o = 60+460 = 520$, $T_i = 530$

i = inside

o = outside

ΔP_s = pressure due to stack effect, inches H_2O, ΔP at height h.—

Building height - 6 floors + 2 basements ≅ 100 ft.

Assume h_{NPL} = 60 feet for wind velocity ≅ 15 mph.

Then:

$$\Delta P_s = (0.00598)(0.075)(32.2)(-55)\left(\frac{530-520}{520}\right)$$

$$= -0.0152 \text{ inches } H_2O.$$

FOR INSIDE TEMP. of 80 F and OUTSIDE TEMP. OF 50 F

$$\Delta P_s = -0.0467 \text{ inches } H_2O,$$

ROGER W. HAINES * CONSULTING ENGINEER * LAGUNA HILLS, CALIFORNIA

Fig. 9-1. An example of a handwritten calculation.

known as experts, our statements may not be taken at face value if they conflict with the experience or vested interests of the client. Additional verification is always desirable and useful.

10

The Oral Presentation

An oral presentation is defined by its name. Quite often, the report writer also will be required to make an oral presentation. This is both an opportunity and a challenge. For those of us who are ambitious, it is an opportunity to test and demonstrate the extent of our knowledge and capability. For all of us, it is an opportunity to explain our reasoning and convince the client of the value of our recommendations. An oral presentation is a challenge in that questions will be asked that require good answers, information will be furnished that was not brought out during the study, and we will be required to prove our analysis and conclusions. Only when we have properly prepared, can we deal adequately and comfortably with the many situations that arise during an oral presentation. Thorough preparation is required because no matter how well the report is written, a poor oral presentation can damage a report's credibility.

THE AUDIENCE

The audience for the oral presentation is the same as that for the report. It includes not only the technical people with whom you have been dealing during the study period, but also the managers of those people and sometimes other groups of operations or managerial staff. In other words, it includes anyone within the client's organization who has an interest in the subject. The larger the organization, the greater will be the diversity of the audience. In a small organization, it may be only two or three people.

THE PRESENTATION

The presenter must tailor the presentation to the audience. Technical issues must be explained in language that all can understand. This is not easy. When you are thoroughly familiar with the project, it is very hard to remember that others may be totally ignorant and must be carefully briefed. Things that are obvious to you are not at all clear to others. On the other hand, it is usually true that some members of the audience will know more about some aspects of the project than the presenter. You must, therefore, avoid condenscension because you are not the only expert present. It is seldom that you will know all the details relating to the situation.

The general rules for public speaking apply: stand erect, speak clearly and not so fast that words are slurred, speak loudly enough to be heard by everyone, and phrase your ideas in complete sentences. If you have a problem with public speaking, join the Toastmasters organization.

The oral presentation should be a summary of the report. It should be based on the summary and recommendations portion of your report and include enough background information so that the audience can understand your conclusions. Never try to present calculations or equipment details unless these data are absolutely essential. There is no surer way to lose your audience. When calculations are involved, it is best to state, "it can be shown..." If someone insists on proof, demonstrate. In the same way, details of a system or equipment specifications should be in the report or its appendix, but need not be described in the oral presentation unless there is a specific need or request.

The purpose of an oral report is to say: This is the background, these are the problems, these are the alternatives with their merits and demerits, and these are our recommendations. Then ask for questions and comments.

VISUAL AIDS

An oral report will not be effective without some kind of visual aid. The simplest form is chalk on a chalkboard or a marking pen on a large, easel-mounted pad. One of these is a necessity when the presentation is impromptu, as sometimes happens. When there is time to prepare, you may use many different techniques: slides, viewgraphs for use on overhead projectors, paintings, models, computer-generated graphics, and video tapes. The report on which the presentation is based is also a visual aid. You can use it by referring to the appropriate page, figure, or table.

Slides. Thirty-five millimeter color slides are a common and effective presentation tool. They are particularly good for showing existing conditions. The basic rule in making slides, or any other visual aids, is that the audience must be able to see and read them easily. Therefore, use heavy line work, large letters, and keep each slide simple (FIG. 10-1). If a complex system must be shown, break it down into separate elements. The idea of simplicity is one of those fundamentals presenters often neglect. This rule is basic to teaching, sales, and propaganda. One idea at a time is all that most of the audience will be able to grasp. Too much too fast, and the audience is lost and seldom regained.

Viewgraphs. These are similar to slides in that they require visibility and simplicity. Viewgraphs can be carefully and artistically drawn, or they can be very simple freehand sketches (FIG. 10-2). Both methods are effective, though one may be better than the other depending on the situation. The overhead projector used to show viewgraphs also provides you with a method of responding to

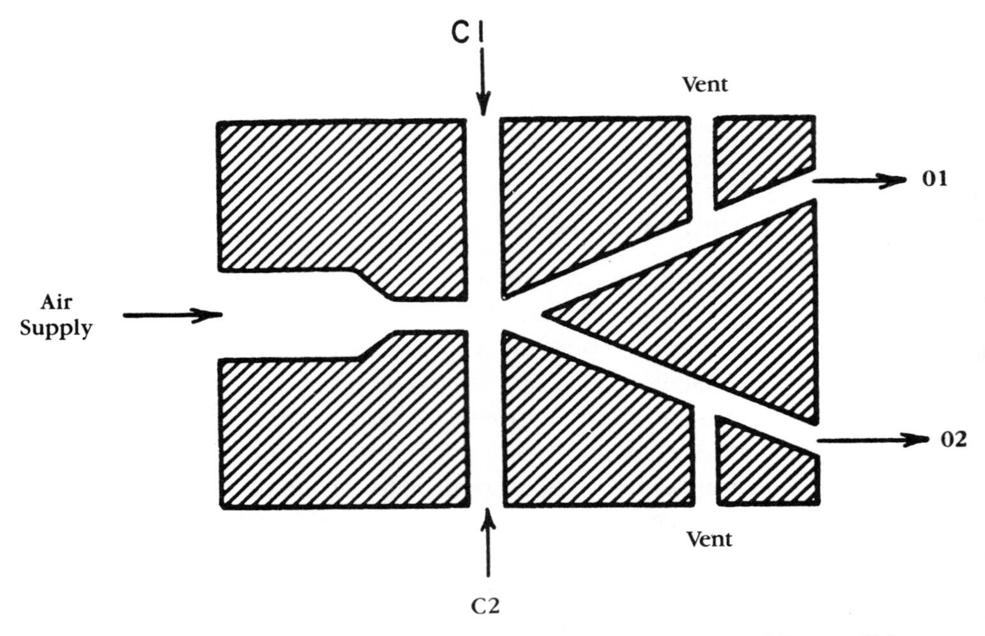

Fig. 10-1. An example of the kind of art appropriate for a 35mm slide.

questions with somewhat better visibility than a chalkboard. Color can be very helpful. Colored pens and colored foil are available for use on viewgraphs.

Painting. When the subject of the report involves extensive remodeling or new construction, it is much easier to visualize the final product when it is shown in a painting. Good commercial artists can provide effective perspective renderings from scale drawings or sketches at a reasonable cost. Sometimes several versions may be shown to help the audience consider alternatives. While paintings are most often used by architects, they can and should be used for any type of engineering project.

Models. A scale model is even more effective than a painting and, usually, more expensive. Models may be constructed so that their disassembly shows internal relationships. The cost of a good model can be more easily justified on a large and complex project.

Computer-generated graphics. Modern computer graphics, especially those generated with a computer aided drafting system, can be very effective. Viewers respond favorably to this new state-of-the-art technology. This does not make the graphics necessarily better or more accurate than older methods, but because it is new technology it may be more effective. The system may not be readily available to you and the cost may be excessive. As with any technique, the use of computer-generated graphics should be suited to the situation.

Video tapes. Video presentations are very popular in this electronic age. There are two problems with the video presentation of a technical report. The cost of producing a professional quality video is not small. It is much more involved than simply standing in front of a camera and talking. There is also the lack of interaction between the presenter and the audience. One of the great values of the oral presentation is the opportunity for members of the audience to ask questions at any time. Frequently, something said in the presentation will trigger a lengthy discussion among the listeners. Thus, even with a professional video, it will be necessary for someone to be present to answer the questions or stop the video to allow discussion.

QUESTIONS AND ANSWERS

Questions are an inevitable part of an oral presentation. How you deal with questions will play a large part in how well the presentation is accepted.

The chairman of the client group may prefer that questions be deferred until the presentation is completed. This is usually the best arrangement for the beginning presenter. When everything is carefully arranged and organized, it is easier to proceed without interruption. This is the proper method for presenting a technical paper. However, from experience I know that the discussion flows more freely and more questions are asked when questions are taken at frequent intervals. As you gain experience and confidence, this will probably become your preferred procedure.

How you answer questions is crucial. Many questions will address issues peripheral to or outside of the scope of the report. To the extent that you know the answers, you should respond, but always qualify the answer as not being within the report's scope of work. Sometimes additional information will be volunteered that could change your conclusions. The appropriate response is always the honest one: "That's new information. I'll have to study it before I can give you an answer."

Sometimes you won't know the answer, despite all of your preparation. Even the most careful study may miss an important point and all of us, at one time or another, have been embarrassed by the question we forgot to ask ourselves. Then the best policy is to admit that you don't know and say you will find out. Trying to bluff an answer is seldom effective and frequently disastrous. Admitting a lack of knowledge may detract temporarily from your image, but following through and obtaining an answer will improve matters. Dishonesty is never a good policy and is always unprofessional.

TIMING AND SCHEDULING

Planning the presentation to fit into a scheduled time is essential. Most members of the audience have a schedule to meet and they will appreciate a presentation that is kept on schedule. In fact, you may lose part of the audience if you talk too long. Or, as has happened, the next group to use the conference room will come knocking on the door and your great conclusion will be lost in the confusion.

Brevity, without omitting any essentials, is the key. This takes a great deal of planning, organization, and experience. For the beginner, even more planning is necessary. Put your watch in plain sight to show the audience that you have a schedule. After a few presentations, you will begin to develop a sense of timing that will help things flow better. As in everything you do, experience helps.

SUMMARY

Not all of us enjoy giving oral presentations. Many engineers are uncomfortable dealing with a group of people. If you are one of these, then oral presentations may not be for you. Again, I suggest that you get help from Toastmasters or a college course in public speaking. For all of us, oral presentations provide a real challenge and stimulus. It is also one of the ways in which reputations are built. Your real competence will show, and be developed by, the give and take discussions that occur during an oral presentation. Any dishonesty or lack of preparation will be evident and help you do better next time. For me, oral presentations are a time of real enjoyment when I can interact with and serve my clients.

11

Litigation

As technical experts, report writers are sometimes called upon to serve as *expert witnesses*. The expert witness is a very special communicator. In the previous chapters, I discussed many aspects of report writing and oral presentation. Expert witnesses work in all of these areas, with a special emphasis on accurate and thorough research. The work includes not only writing a preliminary report, but educating the attorneys for the client in the technical aspects of the case. Often expert witnesses must present the results of their investigations as testimony, both in deposition and during trial.

SPECIAL REQUIREMENTS FOR REPORTS IN LITIGATION

A report that is furnished in connection with a civil or criminal court case will frequently be entered into evidence. I have said a great deal in this book about being careful and thorough when investigating and analyzing the data on which your report is based. This becomes even more critical when you must appear as a witness and defend the report under cross-examination by the opposition attorney with a judge and jury listening. Under the penalty of perjury for an answer you cannot prove, it is sometimes necessary to qualify your answer as an educated guess or an expert opinion.

More often, however, you have the opportunity to bring some order to the conflicting claims of the various parties in the case. A properly organized and documented report can be of great help to the judge and jury. This audience will be even less educated than the typical audience for a technical report, so the language and definitions must be clear enough to educate and clarify. The technical basis must still be defined, however, without talking down to the audience.

If the judge quotes from your report in his decision, then you can be assured that you have done your job well. This is the ultimate criterion.

THE EXPERT WITNESS

The steps involved in serving as an expert witness are discussed below.

Qualifying. An expert witness must be qualified before his opinions can be accepted. He may testify without being qualified but cannot, under those circumstances, offer opinions. A witness must be proven to be an expert by means of education, experience, publications, and peer acceptance. The point to qualification is the establishment of the credibility of the witness. Many times the outcome of the case will depend, to some degree, on the comparative credibility of the expert witnesses.

Formal educational requirements depend on the subject being litigated. A college degree is usually needed to discuss highly technical engineering subjects, such as the design of a structure. For more pragmatic subjects, such as construction practices, experience factors are more important. For highly esoteric subjects, an advanced degree in a specialty may be necessary.

Experience is very important. Experts are always asked about trade practices in their field. Lengthy experience and a few gray hairs are always helpful in establishing credibility. The depth of the expert's experience in relation to the case will be probed carefully by the opposing attorney.

Expertise is very often attributed to those who have a long list of publications. While publication does not assure expertise, experience shows that it creates an impression of expertise. Published writers also study, research, and revise to ensure accuracy. Thus published, writers actually do become more educated and expert. Listing publications is a necessary and valuable part of qualifying a witness.

Peer acceptance relates to how the expert is accepted by others in his specialty. Membership in a technical society and registration as a professional engineer or technician is almost essential. Honors and awards of various kinds are helpful. I am a Fellow of the American Society of Heating, Refrigerating, and Air Conditioning Engineers, a fact that never fails to impress. All of the above criteria are needed to establish a witness as an expert. In the end, however, the real test is in how the witness conducts himself in deposition and testimony.

Deposition. The typical procedure in litigation is to depose a witness prior to the trial. The deposition is conducted by the attorney or attorneys for the opposition, with your client's attorney present to provide guidance. A deposition is a question/answer session during which the witness testifies under oath. The purpose of this exercise is to determine how much the witness knows and to find out what he may testify to at the trial. All or part (or none) of a deposition may be introduced as evidence at the trial.

The opposition attorney may cover a wide range of topics during his questioning. He is trying to determine how well you have prepared, what weaknesses exist in your analysis and, from this, the strength of each side of the dispute. The rules of evidence are stretched somewhat during a deposition. Hypothetical situations are often proposed in an effort to elicit information or even to confuse the witness. The questioner may suggest that something is true that would cause your testimony to be false or questionable. In complicated cases, there may be several questioners and your client's attorney may ask an occasional question for clarification. All of the dialogue is recorded by a court reporter and the witness is required to read, correct, and sign the transcript.

Testimony. A court trial is governed by *rules of evidence*. The nonexpert witness may respond only to questions that he is asked and of which he has direct knowledge. As an expert, you may be given considerable latitude in testimony and may say, "in my opinion... ." You may also answer a question with an explanation, rather than a simple yes or no. In general, witnesses are not allowed in the court during the testimony of other witnesses, but the judge may waive this rule for the expert. The procedure for questioning each witness is outlined below.

- *Direct examination.* The attorney for whom the witness is appearing introduces the witness. If the witness is an expert, his qualifications are developed by appropriate questions and the court is asked to approve him as an expert. The witness is then questioned in accordance with the plan laid down by the attorney. Usually this is discussed beforehand so that the witness knows what to expect.
- *Cross examination.* The opposition attorney then questions the witness. His questioning must be restricted to the areas developed under direct examination. He will endeavor to show the weaknesses in the witness' testimony.
- *Re-direct examination.* The client's attorney may ask additional questions to clarify any problems developed under cross examination. This is particularly advantageous to the expert since he can further explain those answers that may benefit from expansion.
- *Re-cross examination.* This gives the opposition attorney an opportunity to ask additional ques-

tions, especially if the re-direct examination opened up some new areas.

Preparation. When you work with an attorney as an expert, he will give you his rules on how to answer questions in deposition and trial situations. While these vary somewhat among attorneys, there are some basic rules:

- *Take your time in answering.* Wait until the questioner completes the question. Consider what the question asks and the implications. Hesitation will not be held against you. Answer the question as specifically as possible. Sometimes the question will not be clear. You may ask to have it restated. Force the questioner to be specific. Don't help him by asking: do you mean so-and-so.
- *Never volunteer.* As an expert, you can expand as much as you like and there is a tendency to do so. This often leads to trouble. The exception to this rule occurs when a question is asked that cannot be answered unambiguously without some discussion. Use care.
- *If you don't know the answer say so.* Honesty is still the best policy. You will never lose points with this answer. Attempting to answer a question outside the scope of your knowledge or expertise, however, may cause a loss of credibility.
- *If your client's attorney objects, wait to answer.* Don't answer a question unless and until directed by the judge or your client's attorney. The attorney is there to protect your interests, which are the same as those of your client.

ASSISTING THE CLIENT AND CLIENT'S ATTORNEY

While the drama takes place in the courtroom, most of the work is done prior to the trial. Typically, the attorney will contact you about expert testimony. There will always be a preliminary discussion to review the litigation and to determine whether you are a suitable expert. On your side, you need to determine if you can be effective in the situation. You should not take a case that you cannot honestly promote.

When both sides are satisfied, then a fee and schedule may be negotiated. Typically, higher fees are charged for deposition and testimony than for pre-trial study and consultation.

Pre-trial studies and reports are often made, as noted earlier in this chapter. In addition, you will probably spend a great deal of time with the client's attorney educating him or her on the technical aspects of the case. Competent attorneys are very quick to understand and use technical details. You will also be asked to review and comment on the depositions of witnesses for the opposition and note any flaws in their reasoning. These depositions are also useful in discovering areas for further study on your part.

If you are allowed to be present in court during the trial, you should make notes as testimony develops. These notes may be used by your attorney to further examine opposition witnesses, or to develop points during your testimony. The notes should be given to the attorney only during a recess or at the end of the day.

The *discovery* process is used in most cases. Under this rule, either side may ask the other for specific documents relating to the case, including the report of the opposing expert(s). You will be called on to assist in identifying pertinent documents. The general rule is to obtain as much as possible in order to avoid overlooking something. You will be asked to review the documents and use them in your reports and discussions.

SUMMARY

Service as an expert witness is very interesting—and pays well, too. It requires an extraordinary amount of preparation, honesty, patience, care, and professionalism. It can be frustrating and always demands your best effort. There is a high degree of competition involved. Not everyone enjoys it. It is hoped that this chapter has given you some feel for the requirements and whether or not you would be a competent expert.

12

Reporting on Tests and Experiments

Up to this point, I have discussed a general technical report that describes problems and proposes solutions. In this and subsequent chapters, I will consider some special kinds of technical reports. Not all of these deal directly with problem solving, but many common elements are present.

This chapter will discuss the type of report that describes a test or an experiment and its results. An experiment is set up for the purpose of solving a problem or proving or disproving a theory. A test is used to measure the performance of a system or a piece of equipment. The report describes the problem, the experimental procedure, and the results.

THE AUDIENCE

The audience for this type of report is generally a technical one. The report is, of necessity, concerned with technical detail and written at a higher technical level than a report intended for a general audience. The summary still can be written in simple language, however, so that those who do not understand all of the details can readily understand the results.

FORMAT OF THE REPORT

The elements of this type of report should not differ greatly from those of the general report. The arrangement of those elements, however, is traditionally somewhat different. The format includes:

- The purpose of the report and a problem statement
- Proposed approach
- A literature search
- Theory, formulae, and calculations
- Test set-up and procedures to be followed, instrumentation
- Results, data obtained
- Summary, conclusions, and recommendations

PURPOSE AND PROBLEM STATEMENT

This is a statement of the problem or theory that has been investigated or a description of the equipment or system that was tested. The statement should be as simple as possible. Even though many experiments deal with complex subjects, the discipline of writing a simple statement is needed to clarify and focus the problem. This is analogous to the scope of work in the general report and must not only define the problem but limit it. Open-ended statements lead to scattershot approaches to a solution and may easily result in a great deal of wasted effort.

PROPOSED APPROACH TO SOLUTION

Once the problem is properly clarified and stated, an experimental approach can be defined. This may include theoretical analysis, which often involves the use of a computer, or physical testing in the field or a laboratory, or a combination of both.

LITERATURE SEARCH

Most experiments deal with areas where previous experience and data are available. A bibliographical search is often made and the results summarized as part of the report. The experience of the researcher is part of the background, as is informal information from others working in the area. These data are analogous to the existing conditions stated in a general report.

THEORY, FORMULAE, AND CALCULATIONS

Included in this topic are the preliminary analysis of the problem and the derivation or source of the equations used. The algorithms used in the computer programs and the mathematical analysis of the test results should also be included. It is essential that all mathematical analyses be based on generally accepted fundamentals.

TEST SET-UP AND PROCEDURES/INSTRUMENTATION

This is a detailed description of the set-up used for physical testing. It should include figures and/or photos for clarification and a description of the instrumentation used, including quality, accuracy, calibration and recalibration, and frequency of readings. The list of data points and the methods used to record data—manually, with a computer, etc.—should also be included. Standards or codes that prescribe test procedures must also be referenced and followed.

TEST RESULTS AND DATA

The test results must be summarized, analyzed, and presented in a clear, logical way. The presentation might include tables, graphs, diagrams, and other figures, as well as textual discussion. The degree of accuracy should be stated, based on the procedures and instrumentation used. Standards or codes may prescribe the method of presentation. Areas of questionable accuracy should be noted. Data that do not support the original theory must be presented, along with data that are favorable. Improper or biased presentations will decrease credibility.

SUMMARY, CONCLUSIONS, AND RECOMMENDATIONS

This section is analogous to the summary and recommendations section in the general report. The summary is a brief, preferably simple, description of the test procedure and results. The conclusions must always be based on the evidence. The problems with objectivity are the same as those faced when writing the general report, but are now compounded by the fact that someone—the investigator or the theory proponent—usually has a bias in the matter.

The rules to follow are that conclusions must be based on the data and that failure is not necessarily bad. Failure may be good because it may force you to look for other alternatives. Alternatives also may be suggested by the test results. Bending the rules to force a favorable conclusion in an unfavorable situation will almost always lead to future problems. For example, when a manufacturer's laboratory tests a new piece of equipment, modifying or misinterpreting the data to keep the designers or management happy will inevitably lead to dissatisfied customers. The end result will be that no one is happy.

Recommendations may also arise from the results. If the results are not those hoped for, perhaps further work is needed. The report writer should offer suggestions based on the results. Sometimes it is necessary to state that the approach being taken appears to be a dead end. More often the results, even when unfavorable, suggest other approaches that may be better. Many experiments are ongoing so this section describes any further work to be done.

CHAPTER SUMMARY

The report of a test or experiment tends to be more formal than the general report, but it contains many of the same elements and remains an exercise in logical exposition. The need for objective reporting is essential; recognizing that true objectivity is difficult. Failure of the experiment or test should lead to the formulation of alternatives.

13

Reporting Investigations

A report of an investigation is essentially the second section of the general report—existing conditions. In addition, there must be an authorization and scope of work. There also may be some conclusions or recommendations. In many cases, the investigation centers on the degree of compliance or noncompliance on the part of a contractor or supplier with the specifications or instructions issued by a client. The investigation report is common in litigation. It helps form the basis on which a suit is prosecuted or defended.

THE SCOPE OF WORK

The scope of an investigation is usually limited to ascertaining the facts that relate to a specific situation. It is essential that the limits of the investigation be clearly defined in the work statement. As the investigation progresses, it is sometimes necessary to expand the limits and redefine the scope.

Typically, in litigation concerning a construction project, the dispute will involve an interpretation of the specifications or instructions given to the contractor. The investigation must deal with this. In litigation concerning a design, the investigation will be concerned with the adequacy of the design and whether the designer followed typical trade practices and sound engineering principles.

Many investigations are made to provide the basis for a scope of work for a general report. These investigations define the problems. The investigation may be separate from or a part of the general report. Other investigations are inspections. See chapter 14.

INVESTIGATION PROCEDURES

Procedures for the various types of investigations are discussed below.

Field investigations. Field investigations are made for one of several purposes:

- To determine the performance of a system or piece of equipment under as-installed operating conditions
- To determine the conditions existing at a specific point in time. This may be now or at some previous time prior to an event such as a fire, explosion, or modification
- To determine the degree of compliance of an installation with the bid documents

Laboratory investigations. Some tests of materials and equipment cannot be made in the field. This is particularly true for materials, where wear or strength testing must be done. On the other hand, the results of performance tests made in the laboratory will seldom match those made in the field because laboratory conditions are seldom or never encountered in the field. The decision to use the laboratory for investigative purposes must consider the validity and applicability of such investigations.

Computer simulations. In keeping with modern practice, many investigations are made by using computer programs to simulate the conditions and equipment encountered. For hypothetical situations, this will probably provide the most valid results, though manual calculations may also be used. Many people consider the output of the computer program to be definitive, but it is not necessarily accurate. There are some precautions to be taken.

The assumptions made in writing and through use of the computer program must reflect the actual conditions that are being investigated. I have found that small changes in the input data may make large changes in the output data. For example, fan efficiency is one input element in the energy analysis of an air conditioning system. A change of 4 or 5 percent in the efficiency level can substantially change the amount of annual energy consumption. Most of the energy programs assume that the system is always under accurate control. In the real world, however, with typical commercial controls and typical maintenance and operating procedures, the system is seldom controlled accurately. A computer simulation will only accidentally predict the energy consumption of a building under actual operating conditions. Variations of 20 to 30 percent between different programs are common. For these reasons, such programs are normally used only to subjectively compare various system configurations and control strategies.

INVESTIGATION METHODS

The report must describe the investigation procedures and instrumentation used. These data can have a great effect on the accuracy of the results.

Procedures. The methods and procedures used should be described in detail. If codes or standards are followed, these should be referenced and pertinent passages quoted. All procedures should follow accepted engineering practice. Figures, illustrations, or photographs should be used to clarify descriptions. References to previous work may be cited. Regardless of the standing of the expert, the report writer cannot expect unsubstantiated statements to be accepted.

Instrumentation. Laboratory instrumentation must be of the best quality, with good calibration and recalibration procedures followed. Field instrumentation may be of somewhat lower quality, though some modern equipment has been vastly improved. In the field, frequent recalibration is essential. Accuracy must be discussed in the report, since the degree of accuracy determines, to some extent, the degree of the report's validity.

Instrument quality and accuracy can be determined from the manufacturer's specifications. These specifications should stipulate initial and calibrated accuracy, as well as drift with time and response time. If this information is not available, the instrument should be considered inaccurate for reliable test purposes.

RESULTS

Investigation results should be presented in narrative, descriptive form, in tables or graphs, or a combination of both formats. Figures and illustrations should be used. When a computer is used, the computer printout may be included. It is essential that all resultant data be discussed and interpreted in clear, simple language. Answer the question: What do these data mean? Only when the answer is understood by the reader is the report successful. In the case of comparison of an actual situation with a theoretical or expected situation, a graph or table comparing the two will probably be the best method of presentation.

14

Service and Inspection Reports

Service reports are prepared by either in-house service personnel or an outside service organization. These reports describe the service problem and the corrections made. Inspection reports relate to quality control in manufacturing, maintenance, or construction. Most of these reports follow a standard format that is prescribed by company management, but the elements are similar and can be related to the general report structure.

SCOPE OF WORK/PROBLEM STATEMENT

The maintenance person usually has a specific problem to be taken care of, though general or routine maintenance may also be prescribed. This should be addressed in a service report. The problem statement is simple and direct. It is seldom complex. If the statement says unknown problem, then the first step is to define the problem.

Manufacturing inspection reports are usually limited to a specific item or items. The procedures are carefully described and the scope is clearly defined and limited. A maintenance inspection report is a follow-up to maintenance service and evaluates the quality of that service. The scope of this report may be limited or broad. Construction inspection reports prepared by a consulting architect or engineer are now usually called observation reports. They evaluate the contractor's conformance to the design documents. Since a number of observations may be made during construction, the scope of any particular observation may be limited.

EXISTING CONDITIONS

For most service reports, the problem statement defines the existing conditions. For routine maintenance or an unknown or unexpected problem, an investigation of the existing conditions will be necessary and a description should be included in the report.

For a manufacturing inspection, the existing condition report should be a simple pass or fail. It may be necessary to investigate causes of failure if too many items fail. The observation report is, by its nature, a description of existing conditions.

PROCEDURES

For service reports, the procedures section should include a description of the repairs made, with notes regarding parts used, special tools, instrumentation, special procedures, tests, time required, and results.

Manufacturing inspection procedures are usually well-defined. Only deviations from standard procedures will need to be reported. Maintenance inspections and construction observations may require the use of tools and instrumentation. These should be described in the report.

RECOMMENDATIONS

The maintenance procedure may uncover the need for further maintenance or the replacement of equipment. This would become a recommendation in the service report. Recommendations could also include suggestions for changes in maintenance procedures, new tools or instruments, and similar items.

Manufacturing inspection reports could include recommendations for changes in processes or inspection procedures to improve quality and simplify inspection. Maintenance inspection reports also could include recommendations for improving training, standardizing procedures, instrumentation, and tools. Observation inspections usually define who has responsibility for correcting deficiencies. This is, in effect, a recommendation. Observation inspections may also discover the need for changes to the bid documents, which may lead to a recommendation for a change order.

SUMMARY

Service and inspection reports are usually made using standard formats. They are simpler and less formal than general reports. They also contain the same logical elements and should be arranged in a similar manner.

15

Technical Proposals

A technical proposal is an offer to perform a research study or investigation that will usually result in a technical report. In some cases, the work may require a concept or detailed design. The proposal may be made in response to a request for proposal or it may be unsolicited. While the technical proposal is not, in itself, a report, it contains many of the elements that have been discussed throughout this book.

THE REQUEST FOR PROPOSAL

The request for proposal (RFP) is a work statement or scope of work that defines the task to be performed. It often includes a schedule and budget limits. The scope of work was described in chapter 3 and is an essential part of any report. The RFP is written by the client and may be issued through a general advertisement or limited to a select list of proposers. The scope must be limited and carefully defined in order to obtain valid proposals.

An unsolicited proposal is submitted when no request has been made. Typically, such a proposal is made to an agency that is known to have available research funds by a proposer with an idea for a specific research project. In this case, the proposer writes a suggested scope of work and explains other proposal elements, which are described below.

RESPONDING TO THE RFP

The technical proposal is the response to the RFP. It is a sales tool in that its purpose is to convince the client that this particular proposer should be hired to do the work. It includes the scope, technical qualifications of the proposer's personnel, an outline of the proposed procedures, a preliminary schedule for accomplishing the work, and sometimes the fee to be charged. More often, consideration of the fee is postponed until a proposer is selected, at which time a fee is negotiated.

In repeating the scope, the proposer must make sure that he understands it and that the work statement clearly defines and delimits the work. If this is not true, the proposal writer may want to rewrite the scope or contact the client to discuss suggested revisions.

In the outline of proposed procedures, the proposer gets an opportunity to explain why his proposal is superior to all others. In a limited way, this section is similar to the alternatives and recommendations sections of the standard report. For example, you may suggest that certain lines of attack look most promising based on your past experience and study. Even for unusual work projects, there should be some related previous experiences that will be useful.

PRESENTATION OF THE PROPOSAL

All technical proposals are submitted in writing. When competition exists, a few proposers may be asked to come for an interview. This can be similar to an oral presentation or it may be a question-and-answer session. In either case, the rules for oral presentation (chapter 10) apply and visual aids may be used. In many such presentations, the proposer must deal with a committee rather than an individual. It is essential to be well-prepared. That last statement is another of those truisms that would appear to be an insult to the reader, except that I have seen too many cases where the proposer failed to review his proposal, which was written some months previously, before the meeting and was embarrassed.

SUPPORTING DATA

Supporting data in a technical proposal consist of listings of the work experience of the proposer's company and personnel with particular reference to the subject of the RFP.

16

Technical Articles and Papers

Technical articles are intended for publication. Technical papers are usually published after being presented at a technical session. Typically, the paper is part of the proceedings of a technical society, while the article is published in a trade or technical journal. In general, readers expect the paper to be more formal in arrangement and rigorous in logic than the article. The latter is often thought of as being popularized, though this is not necessarily true. While these writings are not reports, they may and often do incorporate some of the elements of a report. Technical papers generally fall into the class of reports on tests, experiments, or investigations, which were discussed in chapters 12 and 13. They will not be discussed further here.

PURPOSE

The technical article is intended to present the writer's ideas. It can be formal or informal. If it is to be effective, it must conform to the rules that have been discussed previously for logical development, clarity, and simplicity. The usual outline starts with a brief summary of the thesis and a statement as to the value of the thesis. This is similar to the scope of work statement in a general report. The summary allows the reader to decide whether or not to read the rest of the article.

The body of the article should be devoted to the development of the thesis, placing emphasis on its application to related problems. This is, to some degree, analogous to the existing conditions and alternatives sections of a general report. Frequently, there will be a need for mathematical exposition. The rules about mathematics in general reports also apply here. For complex but necessary mathematics, it may be desirable to provide an appendix to the paper. In any case, tell the reader what the mathematics mean in relation to the thesis.

The end of the article should restate the thesis and conclusions. This section is equivalent to the summary and recommendations in a general report.

Many articles present a simplified method for calculations in a specific topic area. Discussion should include the limitations and simplifications of these calculation methods so the reader will not use the method improperly. Other articles may describe a solution to a particular problem. In this case, discussion should include suggestions on the applicability of the solution to similar problems. Some articles are simply philosophical. They throw out ideas in an attempt to trigger interest and discussion.

Technical articles are a good place for a beginning writer to develop writing skills. They are usually short, deal with one or two ideas, and can be informal. These articles must be written with care. Use the same audience considerations that you would use in any report. Have something interesting to say, say it simply and logically, and the article should be worth reading.

PREPARATION AND PUBLICATION

Technical articles are generally submitted to periodicals that specialize in the field of interest of the article. There are many such publications, including those dealing with odd and unusual subjects. All publications have minimum requirements for submittals and most have their own style requirements.

All manuscripts must be typed, double spaced, with margins wide enough to allow for editing. While editors often have to deal with manuscripts that have been heavily revised with pencil marks by the authors, they don't really like to work with that kind of material. If you are submitting a first-time article to a publication, make it as neat as possible. Even if it is an article the editor requested from you, make it neat. Be considerate.

Figures should be carefully drawn. They are usually redrawn by the publication's artists to the journal's standards and style. Rough sketches are not suitable, especially when they may be misinterpreted by the editor or artist. The intent of the figure must be explicit.

Most publishers are looking for good articles in their area of interest and will consider most articles submitted. If your article is not accepted, it will do no harm to ask for a critique. Then try again. Not everyone is an instant success!

SUMMARY

When writing a technical paper, follow this rule: Tell the reader what you are going to tell them, then tell them, then tell them what you told them.

17

Technical Books

Writing a technical book is not the same as writing a technical report. Nevertheless, many of the same elements are present. It is still essential that the rules of clear, simple writing and logical development be observed. The preface and introduction should describe the purpose (scope) of the book and the specific problems that will be addressed. The research needed to write a definitive book resembles the data gathering process for a report. The author must bear in mind that the audience will be determined by the technical level and quality of the writing. There also will be alternatives and conclusions in the book, though they may not be expressed as explicitly as in the report.

The discussion in this chapter necessarily will be short. An excellent, detailed treatise on the subject has been written by Daniel N. Fischel.[1]

TYPES OF TECHNICAL BOOKS

Technical books may be divided into three general classifications: professional books, textbooks, and scholarly publications.

Scholarly books are usually, but not necessarily, short. They deal with a limited subject matter and are written at a very high technical level for a rather limited audience. Typically, such books are published by a university press and have a small sale.

Textbooks are written to fit a curriculum in a specific technical subject. They usually include example problems and problems to be solved. In most cases, textbooks are written by a course instructor who feels that existing texts are not adequate. At any time, there may be only a few widely used definitive texts on any subject.

Professional books are written for use by the working professional or technician. They include handbooks, reference books, and cookbooks. Handbooks are written as an education tool, but not as a textbook—there are no problems, for example. They are indexed so the reader can easily reference a particular point of discussion. Many such handbooks have been published, some by individuals, many by technical societies. An example of the latter is *The ASHRAE Handbook*,[2] a four-volume work that contains over 2 million words, which is published by the American Society of Heating, Refrigeration, and Air Conditioning Engineers. It covers the entire field of heating, ventilating, refrigeration, and air conditioning and is revised on a four-year cycle. Most reference books provide data and formulas for use in design, together with the theory behind the data and the derivation of the formulas. The differences between a handbook and a reference book are very minor. Some people include both in the same class. A so-called cookbook provides shortcut data—tables, graphs, nomographs, and formulas. Often, little or no information on the derivation of this data is included.

PLANNING THE BOOK

Most technical books are written because the author believes he has something new to contribute on a subject. Occasionally there will be a gap in the literature—books in the subject area are either nonexistent or out of date. This was the case with the first book I wrote on heating, ventilating, and air conditioning controls. Our committee was writing a chapter on controls for *The ASHRAE Handbook* and I expressed surprise at the lack of bibliographical data. When I said someone ought to write a book, one of the other members suggested I do just that. The results have been very gratifying.

The book must have a plan. It starts with a purpose and scope: what is the intent of the book, why is it needed, can something be contributed to the literature on the subject, who will read it? The answers to these questions will fairly well define the books content, direction, and style.

The next step is to develop an outline. The development must be logical. You must proceed from the known to the unknown one step at a time. This means that you must have the reader in mind: what is his beginning level of knowledge and how far can I take him? Whether you think of it in that way or not, the writing of a technical book is a teaching process. One of the most important rules in teaching is to avoid teaching too much at one time. Only a limited amount can be absorbed by the student. The author's problem is that he knows the subject so well that he is likely to skip some simple step, which may turn out to be a serious problem for the student. Remember that all the steps must be included, even those that are so automatic that you don't realize that you perform them. Take the time to examine all of your thought processes.

After developing an outline and dividing the book into chapters, it is a good idea to write one or two chapters. Good writing is hard work. Don't be afraid to revise. It is helpful to try out your preliminary work on a friend or colleague who is not afraid to criticize. At this point, you may find that what was to be a simple chapter is actually complex. It may be desirable to separate the subject matter into two or more chapters.

SELECTING A PUBLISHER

While it is considered difficult for an unknown author to be accepted by the popular press, this is not true for technical publishers. Most technical publishers will consider an unsolicited submittal.

You can obtain the names of potential publishers by checking the library for similar books. The *Subject Guide to Books in Print*, which is probably available at your library, will give you more information and also help you in your literature search.

When you approach a publisher, send him a copy of your statement of purpose and outline. Include a proposal in which you expand on the purpose, explain the value of the book, describe the potential market, and state how this book differs from, and will be better than, its competitors. If there is a interest, the publisher may send the information to one or two of his experts for review and comment. If your book is found acceptable, you will receive a contract and a schedule for publication. Now all you have to do is finish writing. You may need to try more than one publisher before you find one who has a place in his catalogue for your book, but for a well-planned and needed book the possibilities of publication are very good.

THE SELF-PUBLISHING OPTION

Self-publishing is a reasonable option. The use of modern word processors and laser printers allows writers, with a little effort and care, to produce excellent quality camera-ready copy. There are plenty of print shops that will do the printing and binding. So why bother with a publisher?

Publishing a book involves many more things than printing. To be successful, a book must be sold, which requires advertising, marketing, inventory control, order handling, collections, bookkeeping, reports, taxes, etc. My son has a small business that publishes a legal reference book. Since it requires constant updating, he elected to self-publish. This works well for his situation, but he spends a great deal more time, worry, and effort on sales and distribution than on the actual writing. If you must self-publish, keep this in mind. It is much simpler to let the publisher do the worrying, while you cash the royalty checks when they come in!

PREPARING THE MANUSCRIPT

All that I said about manuscript preparation in chapter 16 becomes even more important when you publish a book. Your publisher will undoubtedly have some special requirements, especially regarding the avoidance of photocopies for figures and tables. These generally do not photograph well. Originals are much better. If your word processor is compatible with that of the publisher, he may ask that you submit copies of the text on disc. Most typesetting is now done by computer.

REFERENCES

1. Daniel N. Fischel, *A Practical Guide to Writing and Publishing Professional Books*, (New York, NY: Van Nostrand Reinhold Company, Inc. 1959).
2. American Society of Heating, Refrigeration and Air Conditioning Engineers, *The ASHRAE Handbook*, Atlanta, GA.

18

Relations with the Client

Every report has a client. The client is not to be confused with the audience, though he or she is usually part of the audience. The client plays an essential role by establishing the scope of work and paying the bill.

The client of a consultant is readily defined by the above criteria. When the report writer is writing for someone within the same company, the client is not so obvious. In the case of an employee, paying the bill means paying the person's salary, as well as the other expenses connected with the report. The scope of work may not be set by the same department that pays the bill. In this case, the client is the person who establishes the scope of work, though he or she also probably has a great deal to say about the status and employment conditions of the report writer. For an employee, the client may also be a committee, which creates still another set of relationships.

THE SELF-EMPLOYED CONSULTANT

The self-employed consultant faces two problems finding potential clients: becoming known and establishing credibility. The consultant becomes known by advertising, word-of-mouth, or publishing. These same things help in the preliminary establishment of credibility.

Advertising in trade publications usually takes the form of a business card, a small box ad that lists the consultant's name and specialties. This is considered professional, anything more is nonprofessional. Direct mail and personal calls are other ways of making yourself known. Government projects are usually advertised and you can respond to these announcements. Being recommended by an old client to a new client is an advantage because it tends to establish a high level of credibility.

My own experience indicates that publishing articles and books is probably the best way to obtain clients and establish credibility. There is something about reading the printed word that convinces others that you are competent. You may not always be convinced of your own competence, but that can be a good thing if it forces you to continually question all that you do.

Having made contact with a potential client, you still have to meet with that client and make a presentation that will establish your credibility. This requires that you do your homework and prepare for the interview. In the interview, you will have to discover the client's needs. This means listening first, then thinking, and finally talking. Reversing this order can only lead to trouble. While you should know as much as possible about the project before the interview, save your prepackaged solution until discussion indicates that it is in fact valid. Most of the time it will need modification. Many times it will become apparent that your skills are not really suitable for the project. Then the choice is to take the job and hire other consultants to help in your weak areas, or to turn down the job. The latter is better than muddling through and performing poorly.

Only after both client and consultant are satisfied with one another is it time to talk about fees. If the consultant's credentials are acceptable and the scope of work is clearly defined, then it should not be difficult to establish a proper fee.

THE CONSULTING OFFICE

When the report writer is part of a larger organization—from two or three to several hundred people—the definition of client begins to blur. Certainly the client is the outside person or group who has employed the consulting office. For the report writer, client may also include the supervisor who will direct the writer's work and review and criticize the report before it leaves the office. When that supervisor has high standards and expectations, the writer's task is harder. I remember vividly considering all sorts of vengeance on the manager who red-penciled (we called it "bleeding all over") my work, which I thought was really pretty good and over which I had labored hard and long. The fact is that I became, in time, an adequate report writer because of that criticism. At least he didn't make as many comments after the first year. The point is that I wrote as much for him as for the real client. This will be true in any office. It is one way in which the beginning writer learns.

THE IN-HOUSE SITUATION

When the report writer is part of a technical group that in turn is part of a larger organization, and some part of that organization is the real client, the relationship is somewhat different. One of the problems that arises is that of conflict. Should the report be shaded towards conclusions that the writer believes are desired by management? The obvious answer is that the true professional will make the report as accurate as possible. That answer may be influenced by economic factors. I can't know how badly you need the paycheck. It is best to remember two things: the best long-term decisions are only made when the facts are presented in an unbiased manner and the individual who writes only what the manager wants to hear will get very little credit for integrity or professionalism.

WHEN THE GOVERNMENT IS THE CLIENT

Government can mean any of the several levels of local, state, or federal offices. Most government work is advertised in order to obtain a wide response. There is usually a policy of spreading the work among as many consultants as possible. Thus, while competence and a good track record are helpful, they by no means ensure that you will get any specific project. Each proposal becomes a brand-new opportunity. In many cases, the government managers will attempt to influence the conclusions of the report to the detriment of everyone concerned. On the other hand, I have dealt with many government representatives who were capable, honest, and hardworking. Take the positive approach. A good, honest report is best.

WHEN THE CLIENT IS A COMMITTEE

The old story that a camel is a horse designed by a committee is still funny and may even be correct some of the time. Still, all report writers are going to have to deal with a committee sooner or later. In this case, the client is also the audience. The secret here is to listen even more carefully. It is sometimes difficult to find out what the majority of the committee really wants. On any committee, there are some who are more technically competent and some who are more vocal. These two subgroups are not necessarily the same. Part of your job is to educate the entire group so that they begin to understand the project and the problems in the same way. You especially have to educate the less competent members so that they can take part in discussions. Even so, it will shortly become apparent that on any committee of equals there are one or two people who are a bit more equal. You will inevitably be guided more by these persons than the others. It is necessary to avoid showing that this is so, however. The need for unbiased reporting is, as always, your first priority.

SUMMARY

The primary requirements in all types of client relationships are: maintaining a professional attitude, unbiased reporting, and listening before you speak.

Index

O

P

Q

R